Student Solutions I

Modeling, Functions, and Graphs
ALGEBRA FOR COLLEGE STUDENTS

KATHERINE FRANKLIN YOSHIWARA
Los Angeles Pierce College

IRVING DROOYAN
Los Angeles Pierce College, Emeritus

Prepared by
STEVEN BLASBERG
West Valley College

PWS PUBLISHING COMPANY

I(T)P   An International Thomson Publishing Company

Boston • New York • London • Bonn • Detroit • Madrid • Melbourne • Mexico City
Paris • Singapore • Tokyo • Toronto • Washington • Albany NY • Belmont CA • Cincinnati OH

PWS PUBLISHING COMPANY
20 Park Plaza, Boston, MA 02116-4324

ISBN: 0-534-13287-1

Printed and bound in the United States of America by Financial Publishing Company.
94 95 96 97 98 --10 9 8 7 6 5 4 3 2 1

Table of Contents

CHAPTER 1

Exercise 1.1

2. 137 is an element of N, W, J, Q and R

4. 2.71828... is an element of H and R (assuming no repeating pattern)

6. $\sqrt{49}$ = 7 is an element of N, W, J, Q and R

8. 0.357357... is an element of Q and R (assuming a repeating pattern)

10. $\sqrt{\dfrac{4}{9}} = \dfrac{2}{3}$ is an element of Q and R   12. $\dfrac{13}{7}$ is an element of Q and R

14. $(-5)^2 = (-5)(-5) = 25$ 16. $-3^4 = -(3)(3)(3)(3) = -81$

18. $(-4)^3 = (-4)(-4)(-4) = -64$ 20. $\dfrac{5(3-5)}{2} - \dfrac{18}{-3} = \dfrac{5(-2)}{2} - \dfrac{18}{-3} = \dfrac{-10}{2} - \dfrac{18}{-3} = -5 - (-6) = 1$

22. $5[3 + 4(6 - 4)] = 5[3 + 4(2)] = 5[3 + 8] = 5[11] = 55$

24. $(8 - 6)[5 + 7(2 - 3)] = 2[5 + 7(-1)] = 2[5 + (-7)] = 2(-2) = -4$

26. $27 \div (3[9 - 3(4 - 2)]) = 27 \div (3[9 - 3(2)]) = 27 \div (3[9 - 6]) = 27 \div (3[3]) = 27 \div 9 = 3$

28. $-3[-2 + (6 - 1)] \div [18 \div (-2)] = -3[-2 + 5] \div (-9) = -3[3] \div (-9) = (-9) \div (-9) = 1$

30. $[-2 + 3(5 - 8)]\cdot[-15 \div (5 - 2)] = [-2 + 3(-3)]\cdot[-15 \div 3] = [-2 + (-9)]\cdot[-5] = (-11)(-5) = 55$

32. $\left[\dfrac{12 + (-2)}{3 + (-8)}\right]\left[\dfrac{6 + (-15)}{8 - 5}\right] = \left[\dfrac{10}{-5}\right]\left[\dfrac{-9}{3}\right] = (-2)(-3) = 6$

34. $\left(7 + 3\left[\dfrac{6 + (-18)}{4 + 2}\right] - 5\right) + 3 = \left(7 + 3\left[\dfrac{-12}{6}\right] - 5\right) + 3 = (7 + (-6) - 5) + 3 = -1$

36. $\dfrac{4\cdot3^2}{6} + (3\cdot4)^2 = \dfrac{4\cdot9}{6} + 12^2 = \dfrac{36}{6} + 144 = 6 + 144 = 150$

38. $\dfrac{3^2\cdot2^2}{4-1} + \dfrac{(-3)(2)^3}{6} = \dfrac{9\cdot4}{3} + \dfrac{(-3)(8)}{6} = \dfrac{36}{3} + \dfrac{-24}{6} = 12 + (-4) = 8$

40. $\dfrac{7^2 - 6^2}{10 + 3} - \dfrac{8^2(-2)}{(-4)^2} = \dfrac{49 - 36}{13} - \dfrac{64(-2)}{16} = \dfrac{13}{13} - \dfrac{-128}{16} = 1 - (-8) = 9$

42. $\dfrac{12 + 3\left(\dfrac{12 - 20}{3^2 - 1}\right)^2 - 1}{-8 + 6\left(\dfrac{12 - 30}{2^4 - 5^2}\right)^2 + 1} = \dfrac{12 + 3\left(\dfrac{-8}{8}\right)^2 - 1}{-8 + 6\left(\dfrac{-18}{16 - 25}\right)^2 + 1} = \dfrac{12 + 3(-1)^2 - 1}{-8 + 6\left(\dfrac{-18}{-9}\right)^2 + 1} = \dfrac{12 + 3(1) - 1}{-8 + 6(4) + 1} = \dfrac{14}{17}$

1

44. $$\frac{62 - 2\left(\frac{4+6}{5}\right)^3 + 8}{3^2 - 3 \cdot 2 + 2^2} = \frac{36 - 2\left(\frac{10}{5}\right)^3 + 8}{9 - 6 + 4} = \frac{36 - 2(8) + 8}{7} = \frac{28}{7} = 4$$

46. $$\frac{R + r}{r} = \frac{12 + 2}{2} = \frac{14}{2} = 7$$

48. $$\frac{a - 4s}{1 - r} = \frac{4 - 4(12)}{1 - 2} = \frac{4 - 48}{-1} = \frac{-44}{-1} = 44$$

50. $$R(1 + at) = 2.5(1 + 0.05(20)) = 2.5(1 + 1) = 5$$

52. $$\frac{1}{2}gt^2 - 12t = \frac{1}{2}(32)3^2 - 12(3) = \frac{1}{2}(32)9 - 36 = 144 - 36 = 108$$

54. $$\frac{Mv^2}{g} = \frac{64(2)^2}{32} = \frac{64(4)}{32} = \frac{256}{32} = 8$$

56. $$\frac{32(V - v)^2}{g} = \frac{32(38.3 - (-6.7))^2}{9.8} = \frac{32(45)^2}{9.8} = \frac{32(2025)}{9.8} = \frac{64{,}800}{9.8} = 6612.245\ldots$$

58. $$ar^{n-1} = -8.0(0.35)^{6-1} = -8.0(0.35)^5 = -8.0(0.0052521875) = -0.042$$

60. $$\frac{a - ar^n}{1 - r} = \frac{6.3 - 6.3(-0.85)^6}{1 - (-0.85)} = \frac{6.3 - 6.3(0.3771495)}{1.85} = \frac{3.9239581}{1.85} = 2.12$$

62. a. $$lwh = (12.3)(4)(7.3) = 359.16 \text{ cubic inches}$$

   b. $$2lw + 2lh + 2wh = 2(6.2)(5.8) + 2(6.2)(2.6) + 2(5.8)(2.6) =$$

   $$71.92 + 32.24 + 30.16 = 134.32 \text{ square feet}$$

64. a. $$\frac{1}{3}\pi r^2 h = \frac{1}{3}(3.14)(4.6)^2(8.1) = \frac{1}{3}(3.14)(21.16)(8.1) = 179.39 \text{ cu. ft.}$$

   b. $$\pi r^2 + \pi rs = 3.14(16)^2 + 3.14(16)(42) = 3.14(256) + 3.14(672) = 2913.12 \text{ sq. cm.}$$

66. a. Let h stand for the height, b_1 and b_2 the bases. Then the area

   is given by $\frac{h}{2}(b_1 + b_2)$.

   b. $$\frac{h}{2}(b_1 + b_2) = \frac{14}{2}(10 + 12) = 7(22) = 154 \text{ square centimeters}$$

68. a. Let r and R stand for the resistances. Then the net resistance

   is given by $\frac{Rr}{R + r}$.

   b. $$\frac{Rr}{R + r} = \frac{10(20)}{10 + 20} = \frac{200}{30} = 6.666\ldots$$

70. a. Let L stand for the length, p the present temperature, and T the temperature when the highway was built. Then the expansion is given by kL(p - T).

   b. $kL(p - T) = 0.000012(1000)(105 - 65) = 0.012(40) = 0.48$ feet

Exercise 1.2

2. $x^2 - 2x + 1$ is a second-degree trinomial; the coefficient of x^2 is 1, the coefficient of x is -2, and 1 is the constant term.

4. $3n + 1$ is a first-degree binomial; the coefficient of n is 3, and 1 is the constant term

6. r^3 is a third-degree monomial; the coefficient of r^3 is 1.

8. $3y^2 + 1$ is a second-degree binomial; the coefficient of y^3 is 3, and 1 is the constant term.

10. b, c, and d are not polynomials

12. $4x^3$ 14. $-11y^2$ 16. $-6z^3 + 5z^2$ 18. $2xy^2 + 3y$ 20. $3r^2 + r$

22. $-2s^2 + s$ 24. $(t^2 - 4t + 1) - (2t^2 - 2) = t^2 - 4t + 1 - 2t^2 + 2 = -t^2 - 4t + 3$

26. $(2u^2 + 4u + 2) - (u^2 - 4u - 1) = 2u^2 + 4u + 2 - u^2 + 4u + 1 = u^2 + 8u + 3$

28. $(4y^4 - 3y^2 - 7) - (6y^2 - y + 2) = 4y^4 - 3y^2 - 7 - 6y^2 + y - 2 = 4y^4 - 9y^2 + y - 9$

30. $(4s^3 - 3s + 2s^2 - 1) - (2s + s^3 - s^2 - 1) = 4s^3 - 3s + 2s^2 - 1 - 2s - s^3 + s^2 + 1 =$

   $3s^3 + 3s^2 - 5s$

32. $7c^2 - 10c + 8$ can be written as $7c^2 - 10c + 8$
   $- \quad -6c^2 - 3c - 2$ $+ \quad 6c^2 + 3c + 2$
                                                      $13c^2 - 7c + 10$

34. $m^2n^2 - 2mn + 7$ can be written as $m^2n^2 - 2mn + 7$
   $- \quad -2m^2n^2 + mn - 3$ $+ \quad 2m^2n^2 - mn + 3$
   $- \quad 3m^2n^2 - 4mn + 2$ $+ \quad -3m^2n^2 + 4mn - 2$
                                                    $mn + 8$

36. $(2t^2 - 3t + 5) + (t^2 + t + 2) - (2t^2 + 3t - 1) = 3t^2 - 2t + 7 - 2t^2 - 3t + 1 = t^2 - 5t + 8$

38. $(2c^2 + 3c + 1) - [(7c^2 + 3c - 2) + (3 - c - 5c^2)] = (2c^2 + 3c + 1) - (2c^2 + 2c + 1) = c$

40. $3x + [2x - x - 4] = 3x + [x - 4] = 4x - 4$

3

42. $5 - [3y + y - 4 - 1] = 5 - [4y - 5] = 5 - 4y + 5 = 10 - 4y$

44. $-x + 3 + [2x - 3 - x - 2] = -x + 3 + x - 5 = -2$

46. $[2y^2 - 4 + y] + [y^2 - 2 - y] = 3y^2 - 6$

48. $-x^2 - 3x - [3x^2 - 6x + x^2 - 2] = -x^2 - 3x - 4x^2 + 6x + 2 = -5x^2 + 3x + 2$

50. $[x - y - x] - (2x - [3x - x + y] + y) = -y - (2x - 3x + x - y + y) = -y - 0 = -y$

52. $-(2y - [2y - 4y + y - 2] + 1) + [2y - 4 + y + 1] = -(2y + y + 2 + 1) + [3y - 3] =$

$-3y - 3 + 3y - 3 = -6$

54. a. For $x = 3$, $2x^3 + x^2 - 3x + 4 = 2(3)^3 + (3)^2 - 3(3) + 4 = 2(27) + 9 - 9 + 4 =$

$54 + 9 - 9 + 4 = 58$

  b. For $x = -3$, $2x^3 + x^2 - 3x + 4 = 2(-3)^3 + (-3)^2 - 3(-3) + 4 = 2(-27) + 9 + 9 + 4 =$

$-54 + 22 = -32$

56. a. For $t = \frac{1}{4}$, $2t^2 - t + 1 = 2\left(\frac{1}{4}\right)^2 - \left(\frac{1}{4}\right) + 1 = 2\left(\frac{1}{16}\right) - \frac{1}{4} + 1 = \frac{1}{8} - \frac{1}{4} + 1 = \frac{7}{8}$

  b. For $t = \frac{-1}{2}$, $2t^2 - t + 1 = 2\left(\frac{-1}{2}\right)^2 - \left(\frac{-1}{2}\right) + 1 = 2\left(\frac{1}{4}\right) + \frac{1}{2} + 1 = \frac{1}{2} + \frac{1}{2} + 1 = 2$

58. a. For $z = 2.1$, $z^3 + 4z - 2 = (2.1)^3 + 4(2.1) - 2 = 9.261 + 8.4 - 2 = 15.661$

  b. For $z = -3.1$, $z^3 + 4z - 2 = (-3.1)^3 + 4(-3.1) - 2 = -29.791 - 12.4 - 2 = -44.191$

60. a. For $a = -1$, $a^5 - a^4 = (-1)^5 - (-1)^4 = -1 - 1 = -2$

  b. For $a = -2$, $a^5 - a^4 = (-2)^5 - (-2)^4 = -32 - 16 = -48$

62. a. $-4.9t^2 + 10,000t + 46$

  b. For $t = 4$, $-4.9(4)^2 + 10,000(4) + 46 = -78.4 + 40,000 + 46 = 39,967.6$ meters;

  for $t = 10$, $-4.9(10)^2 + 10,000(10) + 46 = -490 + 100,000 + 46 = 99,556$ meters.

64. a. If the pool is w feet wide and h feet deep, the surface area is given by

  $2w^2 + wh + wh + 2wh + 2wh$ or $2w^2 + 6wh$.

  b. For $w = 12$ and $h = 6$, $2(12)^2 + 6(12)(6) = 288 + 432 = 720$ square feet $=$

  $\frac{720}{9}$ or 80 square yards (9 square feet = 1 square yard).

66. a. The cost is given by $1.20(2w^2) + 0.80(6wh) = 2.4w^2 + 4.8wh$

4

b. Total cost is $2.4(12)^2 + 4.8(12)(6) = 345.6 + 345.6 = \$691.20$.

68. a. Profit $= (8.8x - 0.004x^2) - (4x + 200) = -0.004x^2 + 4.8x - 200$

    b. For $x = 200$ lb, profit $= -0.004(200)^2 + 4.8(200) - 200 = -160 + 960 - 200 = \$600$

70. a. Profit $= (200x - 0.3x^2) - (120x + 80) = -0.3x^2 + 80x - 80$

    b. For $x = 200$ mowers, profit $= -0.3(200)^2 + 80(200) - 80 =$

        $-12,000 + 16,000 - 80 = \$3,920$.

72. a. The area is given by $(2x)(x) + \dfrac{1}{2}\left[\pi\left(\dfrac{x}{2}\right)^2\right]$ or $2x^2 + \dfrac{\pi x^2}{8}$.

    b. Area is $2(3)^2 + \dfrac{\pi(3)^2}{8} = 18 + \dfrac{9\pi}{8}$ or approximately 21.53 square feet

74. a. If r is the radius and L is the length, the volume is given by $\pi r^2 L + \dfrac{4}{3}\pi r^3$.

    b. If $r = \dfrac{1}{4}L$, the volume is $\pi\left(\dfrac{1}{4}L\right)^2 L + \dfrac{4}{3}\pi\left(\dfrac{1}{4}L\right)^3 = \dfrac{1}{4}\pi L^3 + \dfrac{1}{48}\pi L^3 = \dfrac{13}{48}\pi L^3$,

        which can also be written as $\dfrac{52}{3}\pi r^3$, since $L = 4r$.

Exercise 1.3

2. $(yz^4)^2 = y^2(z^4)^2 = y^2 z^8$ 4. $(3x^2yz^2)^3 = 3^3(x^2)^3 y^3(z^2)^3 = 27x^6y^3z^6$

6. $(-3a^2bc^3)^3 = (-3)^3(a^2)^3 b^3(c^3)^3 = -27a^6b^3c^9$ 8. $(4c^3)(2c) = 8c^4$

10. $(-6r^2s^2)(5rs^3) = -6\cdot5 r^2 r s^2 s^3 = -30r^3s^5$

12. $-5(ab^3)(-3a^2bc) = -5(-3)a a^2 b^3 bc = 15a^3b^4c$

14. $(-5mn)(2m^2n)(-n^3) = -5(-1)m m^2 n n n^3 = 5m^3n^5$

16. $(-3xy)(2xz^4)(3x^3y^2z) = -3(2)(3)x x x^3 y y^2 z^4 z = -18x^5y^3z^5$

18. $-a^2(ab^2)(2a)(-3b^2) = (-1)(2)(-3)a^2 a a b^2 b^2 = 6a^4b^4$

20. $(2xy^3)^4 + (3xy)^2 = 2^4x^4y^{12} + 3^2x^2y^2 = 16x^4y^{12} + 9x^2y^2$

22. $(3r^2st)^3 - (4r^2s^2t^2)^2 = 3^3r^6s^3t^3 - 4^2r^4s^4t^4 = 27r^6s^3t^3 - 16r^4s^4t^4$

24. $(m^2n)(mn)^2 + (mn^2)^2 = m^2n m^2n^2 + m^2n^4 = m^4n^3 + m^2n^4$

26. $(xy)^2 + (-x^2y)^2(-xy^2) = x^2y^2 + (-1)^2x^4y^2(-1)xy^2 = x^2y^2 - x^5y^4$

28. $3(x^2y)^2 + x(x^2y) - x^2(x^2y^2) = 3x^4y^2 + x^3y - x^4y^2 = 2x^4y^2 + x^3y$

30. $2u^2(v^3) + 4v(uv)^2 - uv(uv^2) = 2u^2v^3 + 4vu^2v^2 - u^2v^3 = 5u^2v^3$

32. $3x(2x + y) = 3x(2x) + 3x(y) = 6x^2 + 3xy$

34. $-2y(y^2 - 3y + 2) = -2y(y^2) - (-2y)(3y) + (-2y)2 = -2y^3 + 6y^2 - 4y$

36. $ab^3(-a2b^2 + 4ab - 3) = ab^3(-a^2b^2) + ab^3(4ab) - ab^3(3) = -a^3b^5 + 4a^2b^4 - 3ab^3$

38. $5x^2y^2(3x^4y^2 + 3x^2y - xy^6) = 5x^2y^2(3x^4y^2) + 5x^2y^2(3x^2y) - 5x^2y^2(xy^6) =$

    $15x^6y^4 + 15x^4y^3 - 5x^3y^8$

40. $(t + 4)(t^2 - t - 1) = t(t^2 - t - 1) + 4(t^2 - t - 1) = t^3 - t^2 - t + 4t^2 - 4t - 4 = t^3 + 3t^2 - 5t - 4$

42. $(x - 7)(x^2 - 3x + 1) = x(x^2 - 3x + 1) - 7(x^2 - 3x + 1) = x^3 - 3x^2 + x - 7x^2 + 21x - 7 =$

    $x^3 - 10x^2 + 22x - 7$

44. $(y - 2)(y + 2)(y + 4) = [y(y + 2) - 2(y + 2)](y + 4) = (y^2 + 2y - 2y - 4)(y + 4) =$

    $(y^2 - 4)(y + 4) = y^2(y + 4) - 4(y + 4) = y^3 + 4y^2 - 4y - 16$

46. $(z - 5)(z + 6)(z - 1) = (z - 5)(z^2 + 6z - z - 6) = (z - 5)(z^2 + 5z - 6) =$

    $z(z^2 + 5z - 6) - 5(z^2 + 5z - 6) = z^3 + 5z^2 - 6z - 5z^2 - 25z + 30 = z^3 - 31z + 30$

48. $(3x - 2)(4x^2 + x - 2) = 3x(4x^2 + x - 2) - 2(4x^2 + x - 2) =$

    $12x^3 + 3x^2 - 6x - 8x^2 - 2x + 4 = 12x^3 - 5x^2 - 8x + 4$

50. $(b^2 - 3b + 5)(2b^2 - b + 1) = b^2(2b^2 - b + 1) - 3b(2b^2 - b + 1) + 5(2b^2 - b + 1) =$

    $2b^4 - b^3 + b^2 - 6b^3 + 3b^2 - 3b + 10b^2 - 5b + 5 = 2b^4 - 7b^3 + 14b^2 - 8b + 5$

52. $(r - 1)(r - 6) = r^2 - 6r - r + 6 = r^2 - 7r + 6$

54. $(z - 3)(z + 5) = z^2 - 3z + 5z - 15 = z^2 + 2z - 15$

56. $(3t - 1)(2t + 1) = 6t^2 - 2t + 3t - 1 = 6t^2 + t - 1$

58. $(2z - w)(3z + 5w) = 6z^2 - 3wz + 10wz - 5w^2 = 6z^2 + 7wz - 5w^2$

60. $(3a + 5b)(3a + 4b) = 9a^2 + 15ab + 12ab + 20b^2 = 9a^2 + 27ab + 20b^2$

62. $(2x - 3z)(2x + 3z) = 4x^2 - 6xz + 6xz - 9z^2 = 4x^2 - 9z^2$

64. $(s^2 - 5t^2)(3s^2 + 2t^2) = 3s^4 - 15s^2t^2 + 2s^2t^2 - 10t^4 = 3s^4 - 13s^2t^2 - 10t^4$

66. $3[2a - (a + 1) + 3] = 3[2a - a - 1 + 3] = 3[a + 2] = 3a + 6$

68. $-2a[3a + (a - 3) - (2a + 1)] = -2a[3a + a - 3 - 2a - 1] = -2a[2a - 4] = -4a^2 + 8a$

70. $-4(4 - [3 - 2(x - 1) + x] + x) = -4(4 - [3 - 2x + 2 + x] + x) = -4(4 - [5 - x] + x) =$

$-4(4 - 5 + x + x) = -4(-1 + 2x) = 4 - 8x$

72. $x(4 - 2[3 - 4(x + 1)] -x) = x(4 - 2[3 - 4x - 4] - x) = x(4 - 2[-1 - 4x] - x) =$

$x(4 + 2 + 8x - x) = x(6 + 7x) = 6x + 7x^2$

74. $-3[2x^2 - 3(x - 2)(x + 3) + 3x] = -3[2x^2 - 3(x^2 + x - 6) + 3x] =$

$-3[2x^2 - 3x^2 - 3x + 18 + 3x] = -3[-x^2 + 18] = 3x^2 - 54$

76. $x(x[x(x + 3) - 2] - 3) - 5 = x(x[x^2 + 3x - 2] - 3) - 5 = x(x^3 + 3x^2 - 2x - 3) - 5 =$

$x^4 + 3x^3 - 2x^2 - 3x - 5$

78. a. $\frac{1}{6}n(n - 1)(n - 2) = \frac{1}{6}n(n^2 - 3n + 2) = \frac{n^3 - 3n^2 + 2n}{6}$

b. $\frac{12^3 - 3(12)^2 + 2(12)}{6} = \frac{1728 - 3(144) + 24}{6} = \frac{1320}{6} = 220$

80. a. $x(800 - 4x) = 800x - 4x^2$

b. At $x = \$45$, the revenue is $45(800 - 4(45)) = 45(800 - 180) = \$27,900$.

82. a. Price of a room $= 24 + 2x$; number of rooms occupied $= 32 - x$

b. Total income received is $(24 + 2x)(32 - x) = 768 + 40x - 2x^2$

c. At \$30 for a room, $x = 3$ and income $= 768 + 40(3) - 2(3)^2 = 768 + 120 - 18 =$

$\$870$; at \$20, $x = -2$ and income $= 768 + 40(-2) - 2(-2)^2 = 768 - 80 - 8 = \$680$.

84. a. Size of the group $= 20 + x$; price per person $= 600 - 10x$.

b. Total income $= (20 + x)(600 - 10x) = 12,000 + 400x - 10x^2$

c. For 25 people, $x = 5$, and income $= 12,000 + 400(5) - 10(5)^2 =$

$12,000 + 2,000 - 250 = \$13,750$; for 30 people, $x = 10$, and income $=$

$12,000 + 400(10) - 10(10)^2 = 12,000 + 4,000 - 1,000 = \$15,000$.

86. $b^n b^{2n+1} = b^{n+2n+1} = b^{3n+1}$ 88. $a^{2n-2} a^{n+3} = a^{2n-2+n+3} = a^{3n+1}$

90. $(xy^{3n})^2 = x^2(y^{3n})^2 = x^2 y^{6n}$ 92. $(x^{n-2} y^{2n+1})^2 = (x^{n-2})^2 (y^{2n+1})^2 = x^{2n-4} y^{4n+2}$

94. $3t^n(2t^n + 3) = 3(2)t^n t^n + 3(3)t^n = 6t^{2n} + 9t^n$

96. $b^{2n+2}(b^{n-1} + b^n) = b^{2n+2} b^{n-1} + b^{2n+2} b^n = b^{3n+1} + b^{3n+2}$

98. $(a^n - 3)(a^n + 2) = a^n a^n - 3a^n + 2a^n - 6 = a^{2n} - a^n - 6$

100. $(a^{2n} - 2b^n)(a^{3n} + b^{2n}) = a^{2n}a^{3n} - 2b^na^{3n} + a^{2n}b^{2n} - 2b^nb^{2n} =$

$a^{5n} - 2a^{3n}b^n + a^{2n}b^{2n} - 2b^{3n}$

102. $(x - a)^2 = (x - a)(x - a) = xx - ax - ax + aa = x^2 - 2ax + a^2$

104. $(x + a)(x + b) = xx + ax + bx + ab = x^2 + (a + b)x + ab$

106. $(x - a)(x^2 + ax + a^2) = x^3 + ax^2 + a^2x - ax^2 - a^2x - a^3 = x^3 - a^3$

Exercise 1.4

2. $3x^2y + 6xy = 3xy(x + 2)$ 4. $2x^4 - 4x^2 + 8x = 2x(x^3 - 2x + 4)$

6. $2x^2y^2 - 3xy + 5x^2 = x(2xy^2 - 3y + 5x)$

8. $6x^3y - 6xy^3 + 12x^2y^2 = 6xy(x^2 - y^2 + 2xy)$

10. $14xy^4z^3 + 21x^2y^3z^2 - 28x^3y^2z^5 = 7xy^2z^2(2y^2z + 3xy - 4x^2z^3)$

12. $b(a - 2) + a(a - 2) = (b + a)(a - 2)$ 14. $2x(x + 3) - y(x + 3) = (2x - y)(x + 3)$

16. $6(x + 1) - 3x(x + 1)^2 = (6 - 3x(x + 1))(x + 1) = (6 - 3x - 3x^2)(x + 1)$

18. $x^2(x + 3)^3 - x(x + 3)^2 = x(x + 3)^2(x(x + 3) - 1) = x(x + 3)^2(x^2 + 3x - 1)$

20. $(x + 2)^2(x - 1) - (x + 2)^2 = (x + 2)^2((x - 1) - 1) = (x + 2)^2(x-2)$

22. $3(x + 2)^2(x - 4) + 6(x + 2)(x + 1)^2 = 3(x + 2)((x + 2)(x - 4) + 2(x + 1)^2) =$

$3(x + 2)(x^2 - 2x - 8 + 2x^2 + 4x + 2) = 3(x + 2)(3x^2 + 2x - 6)$

24. $2a - b = -(b - 2a)$ 26. $-6x - 9 = -3(2x + 3)$ 28. $-a^2 + ab = -a(a - b)$

30. $3x + 3y - 2z = -(-3x - 3y + 2z)$ 32. $x^2 + 5x + 4 = (x + 4)(x + 1)$

34. $y^2 - 7y + 10 = (y - 5)(y - 2)$ 36. $x^2 - 15 - 2x = x^2 - 2x - 15 = (x - 5)(x + 3)$

38. $3x^2 - 7x + 2 = (3x - 1)(x - 2)$ 40. $1 - 5x + 6x^2 = (2x - 1)(3x - 1)$

42. $10y^2 - 3y - 18 = (5y + 6)(2y - 3)$ 44. $8u^2 - 3 + 5u = 8u^2 + 5u - 3 = (8u - 3)(u + 1)$

46. $24x^2 - 29x + 5 = (24x - 5)(x - 1)$

48. $-30a + 72a^2 - 25 = 72a^2 - 30a - 25 = (12a + 5)(6a - 5)$

50. $39x + 80x^2 - 20 = 80x^2 + 39x - 20 = (16x - 5)(5x + 4)$

52. $48t^2 - 122t + 39 = (6t - 13)(8t - 3)$ 54. $9x^2 + 9ax - 10a^2 = (3x + 5a)(3x - 2a)$

56. $12x^2 + 7xy - 12y^2 = (4x - 3y)(3x + 4y)$

58. $24u^2 - 20v^2 + 17uv = 24u^2 + 17uv - 20v^2 = (8u - 5v)(3u + 4v)$

60. $24a^2 - 15b^2 - 2ab = 24a^2 - 2ab - 15b^2 = (6a - 5b)(4a + 3b)$

62. $12a^2b^2 - ab - 20 = 12(ab)^2 - (ab) - 20 = (4ab + 5)(3ab - 4)$

64. $54x^2y^2 + 3xy - 2 = 54(xy)^2 + 3(xy) - 2 = (9xy + 2)(6xy - 1)$

66. $26a^2z^2 - 24 + 23az = 26(az)^2 + 23(az) - 24 = (13az - 8)(2az + 3)$

68. $ax^2 + x + a^2x + a = x(ax + 1) + a(ax + 1) = (x + a)(ax + 1)$

70. $x^3 - x^2y + xy - y^2 = x^2(x - y) + y(x - y) = (x^2 + y)(x - y)$

72. $5xz + y - 5yz - x = 5xz - 5yz - x + y = 5z(x - y) - 1(x - y) = (5z - 1)(x - y)$

74. $2a^2 + 3a - 2ab - 3b = a(2a + 3) - b(2a + 3) = (a - b)(2a + 3)$

76. $3x^2 - 3x + 2xy - 2y = 3x(x - 1) + 2y(x - 1) = (3x + 2y)(x - 1)$

78. $12 + 3x^2 - 4y^3 - x^2y^3 = 12 - 4y^3 + 3x^2 - x^2y^3 = 4(3 - y^3) + x^2(3 - y^3) = (4 + x^2)(3 - y^3)$

80. $x^3 + 9 + 3x^2 + 3x = x^3 + 3x^2 + 3x + 9 = x^2(x + 3) + 3(x + 3) = (x^2 + 3)(x + 3)$

82. $2x^3 + 14 + 7x^2 + 4x = 2x^3 + 7x^2 + 4x + 14 = x^2(2x + 7) + 2(2x + 7) = (x^2 + 2)(2x + 7)$

84. $2x^2y + 6xy - 20y = 2y(x^2 + 3x - 10) = 2y(x + 5)(x - 2)$

86. $2a^3 - 8a^2 - 10a = 2a(a^2 - 4a - 5) = 2a(a - 5)(a + 1)$

88. $20a^2 + 60ab + 45b^2 = 5(4a^2 + 12ab + 9b^2) = 5(2a + 3b)(2a + 3b)$

90. $9x^3y + 18x^2y^2 + 8xy^3 = xy(9x^2 + 18xy + 8y^2) = xy(3x + 4y)(3x + 2y)$

92. $9u^2v^3 + 12uv^2 - 12v = 3v(3(uv)^2 + 4(uv) - 4) = 3v(3uv - 2)(uv + 2)$

94. $16s^3t^3 - 16s^2t^2 - 12st = 4st(4(st)^2 - 4(st) - 3) = 4st(2st - 3)(2st + 1)$

96. $x^{4n} + x^{2n} = x^{2n}(x^{2n} + 1)$ 98. $4y^{4n} + 3y^{3n} + 2y^{2n} = y^{2n}(4y^{2n} + 3y^n + 2)$

100. $6x^{n+2} - 3x^{n+1} - 3x^n = 3x^n(2x^2 - x - 1) = 3x^n(2x + 1)(x - 1)$

Exercise 1.5

2. $(y - 4)^2 = y^2 - 2(4)(y) + 4^2 = y^2 - 8y + 16$

4. $(3x + 2)^2 = (3x)^2 + 2(2)(3x) + 2^2 = 9x^2 + 12x + 4$

6. $(x - 7)(x + 7) = x^2 - 7^2 = x^2 - 49$ 8. $(2x + a)(2x - a) = (2x)^2 - a^2 = 4x^2 - a^2$

10. $(4u + 5v)^2 = (4u)^2 + 2(4u)(5v) + (5v)^2 = 16u^2 + 40uv + 25v^2$

12. $(7yz - 2)^2 = (7yz)^2 - 2(2)(7yz) + 2^2 = 49y^2z^2 - 28yz + 4$

14. $3[2x + (x + 2)^2] = 3[2x + x^2 + 2(2)(x) + 2^2] = 3[x^2 + 6x + 4] = 3x^2 + 18x + 12$

16. $-2x + x[3 - (x + 4)^2] = -2x + x[3 - (x^2 + 8x + 16)] = -2x + x[3 - x^2 - 8x - 16] =$

$-2x + x[-x^2 - 8x - 13] = -2x + -x^3 - 8x^2 - 13x = -x^3 - 8x^2 - 15x$

18. $-x[2x - (2x + 1)^2 + 3] = -x[2x - (4x^2 + 4x + 1) + 3] = -x[2x - 4x^2 - 4x - 1 + 3] =$

$-x[-4x^2 - 2x + 2] = 4x^3 + 2x^2 - 2x$

20. $x^2 - 36 = x^2 - 6^2 = (x + 6)(x - 6)$

22. $x^2 + 26x + 169 = x^2 + 2(13)x + 13^2 = (x + 13)^2$

24. $9x^2 - y^2 = (3x)^2 - y^2 = (3x + y)(3x - y)$

26. $4y^2 + 4y + 1 = (2y)^2 + 2(1)(2y) + 1^2 = (2y + 1)^2$

28. $16s^2 - 56st + 49t^2 = (4s)^2 - 2(4s)(7t) + (7t)^2 = (4s - 7t)^2$

30. $16a^2 - 9b^2 = (4a)^2 - (3b)^2 = (4a + 3b)(4a - 3b)$

32. $x^2y^2 - 64 = (xy)^2 - 8^2 = (xy + 8)(xy - 8)$

34. $4x^2y^2 + 12xy + 9 = (2xy)^2 + 2(2xy)(3) + 3^2 = (2xy + 3)^2$

36. $64x^2y^2 - 1 = (8xy)^2 - 1^2 = (8xy + 1)(8xy - 1)$

38. $x^2 - (y - 3)^2 = (x + (y - 3))(x - (y - 3)) = (x + y - 3)(x - y + 3)$

40. $x^2 - 6x + 9 - y^2 = (x - 3)^2 - y^2 = (x - 3 + y)(x - 3 - y)$

42. $y^2 - x^2 + 4x - 4 = y^2 - (x^2 - 4x + 4) = y^2 - (x - 2)^2 = (y + x - 2)(y - x + 2)$

44. $9x^2 - 6x + 1 - 9y^2 = (3x - 1)^2 - 9y^2 = (3x - 1 + 3y)(3x - 1 - 3y)$

46. $(x + 2)(x^2 - 2x + 4) = (x + 2)(x^2 - 2x + 2^2) = x^3 + 2^3 = x^3 + 8$

48. $(3x - 1)(9x^2 + 3x + 1) = (3x - 1)((3x)^2 + (3x)(1) + 1^2) = (3x)^3 - 1^3 = 27x^3 - 1$

50. $(2a + 3b)(4a^2 - 6ab + 9b^2) = (2a + 3b)((2a)^2 - (2a)(3b) + (3b)^2) = (2a)^3 + (3b)^3 =$

$8a^3 + 27b^3$

52. $y^3 - 1 = y^3 - 1^3 = (y - 1)(y^2 + 1y + 1^2) = (y - 1)(y^2 + y + 1)$

54. $y^3 + (3x)^3 = (y + 3x)(y^2 - y(3x) + (3x)^2) = (y + 3x)(y^2 - 3xy + 9x^2)$

56. $27a^3 + b^3 = (3a)^3 + b^3 = (3a + b)((3a)^2 - (3a)b + b^2) = (3a + b)(9a^2 - 3ab + b^2)$

10

58. $8 + x^3y^3 = 2^3 + (xy)^3 = (2 + xy)(2^2 - 2xy + (xy)^2) = (2 + xy)(4 - 2xy + x^2y^2)$

60. $8a^3 - 125b^3 = (2a)^3 - (5b)^3 = (2a - 5b)((2a)^2 + 2a(5b) + (5b)^2) =$

    $(2a - 5b)(4a^2 + 10ab + 25b^2)$

62. $64a^3b^3 + 1 = (4ab)^3 + 1^3 = (4ab + 1)((4ab)^2 - 4ab(1) + 1^2) = (4ab + 1)(16a^2b^2 - 4ab + 1)$

64. $(x + y)^3 - z^3 = (x + y - z)((x + y)^2 + (x + y)z + z^2) =$

    $(x + y - z)(x^2 + 2xy + y^2 + xz + yz + z^2)$

66. $x^3 + (x - 2y)^3 = (x + x - 2y)(x^2 - x(x - 2y) + (x - 2y)^2) =$

    $(2x - 2y)(x^2 - x^2 + 2xy + x^2 - 4xy + 4y^2) = 2(x - y)(x^2 - 2xy + 4y^2)$

68. $(2y - 1)^3 + (y - 1)^3 = (2y - 1 + y - 1)((2y - 1)^2 - (2y - 1)(y - 1) + (y - 1)^2) =$

    $(3y - 2)(4y^2 - 4y + 1 - (2y^2 - 3y + 1) + y^2 - 2y + 1) = (3y - 2)(3y^2 - 3y + 1)$

70. $y^4 - 49 = (y^2)^2 - 7^2 = (y^2 + 7)(y^2 - 7)$ 72. $a^4 - 5a^2 + 6 = (a^2 - 3)(a^2 - 2)$

74. $4x^4 - 11x - 3 = (x^2 - 3)(4x^2 + 1)$

76. $x^4 - 81 = (x^2)^2 - 9^2 = (x^2 + 9)(x^2 - 3^2) = (x^2 + 9)(x + 3)(x - 3)$

78. $x^6 - 6x^3 - 27 = (x^3 - 9)(x^3 + 3)$

80. $u^8 - 13u^4 + 36 = (u^4 - 3^2)(u^4 - 2^2) = (u^2 + 3)(u^2 - 3)(u^2 + 2)(u^2 - 2)$

82. $x^3 - 4x^3y^2 = x^3(1 - 4y^2) = x^3(1 - (2y)^2) = x^3(1 + 2y)(1 - 2y)$

84. $x^3y - xy^3 = xy(x^2 - y^2) = xy(x + y)(x - y)$

86. $2a^3 - 54a^3b^9 = 2a^3(1 - (3b^3)^3) = 2a^3(1 - 3b^3)(1 + 3b^3 + 9b^6)$

88. $9a^3b^6 + 3a^3b^4 - 6a^3b^2 = 3a^3b^2(3b^4 + b^2 - 2) = 3a^3b^2(3b^2 - 2)(b^2 + 1)$

90. $6ax^5 + 9a^3x^3 - 6a^5x = 3ax(2x^4 + 3a^2x^2 - 2a^4) = 3ax(2x^2 - a^2)(x^2 + 2a^2)$

92. $4x^8 - 30x^5 - 54x^2 = 2x^2(2x^6 - 15x^3 - 27) = 2x^2(2x^3 + 3)(x^3 - 9)$

94. $x^9 - y^9 = (x^3 - y^3)(x^6 + x^3y^3 + y^6) = (x - y)(x^2 + xy + y^2)(x^6 + x^3y^3 + y^6)$

96.   a. $800(1 + r)^2, 800(1 + r)^3, 800(1 + r)^4$

       b. $800 + 1600r + 800r^2, 800 + 2400r + 2400r^2 + 800r^3, 800 + 3200r + 4800r^2 +$

             $3200r^3 + 800r^4$

       c. $800(1.12)^2 = \$1{,}003.52, 800(1.12)^3 = \$1{,}123.94, 800(1.12)^4 = \$1{,}258.82$

11

98. a. length and width are both 20 - 2x, height is x

b. volume = $x(20 - 2x)^2 = x(400 - 80x + 4x^2) = 4x^3 - 80x^2 + 400x$

c. surface area = $4x(20 - 2x) + (20 - 2x)^2 = 80x - 8x^2 + 400 - 80x + 4x^2 =$

$-4x^2 + 400$

100. a. If r is the smallest radius, the radii are r, $r + \frac{1}{2}$, and $r + 1$ feet.

b. Volumes are $\pi r^2(2)$, $\pi(r + \frac{1}{2})^2(3)$, and $\pi(r + 1)^2(4)$, or $2\pi r^2$, $3\pi(r + \frac{1}{2})^2$, and

$4\pi(r + 1)^2$

c. Total volume of a set = $2\pi r^2 + 3\pi(r^2 + r + \frac{1}{4}) + 4\pi(r^2 + 2r + 1) =$
$9\pi r^2 + 11\pi r + 4.75\pi$

102. a. No (see b. and c.)   b. $x^2 + y^2$ does not factor

c. $(x + y)^2 = x^2 + 2xy + y^2 \neq x^2 + y^2$

104. a. No (see b. and c.)   b. $x^3 + y^3 = (x + y)(x^2 - xy + y^2)$

c. $(x + y)^3 = x^3 + 3x^2y + 3xy^2 + y^3 \neq x^3 + y^3$

106. a. $x^2 - y^2$ b. $(x + y)(x - y)$

c.

108. a. $\pi x^2 - \pi y^2$ b. $\pi(x + y)(x - y)$ c. 66π square inches

CHAPTER 2

Exercise 2.1

2. $\dfrac{3}{4}$

4. $-\dfrac{4}{5}$

6. $\dfrac{x}{2y}$, $y \neq 0$

8. $-\dfrac{x+3}{x}$, $x \neq 0$

10. $-\dfrac{3y-2}{4y-1}$, $y \neq \dfrac{1}{4}$

12. $-\dfrac{3y^2+8}{12y^2-3}$, $y \neq \pm\dfrac{1}{2}$

14. $\dfrac{-3}{2-x} = \dfrac{3}{-(2-x)} = \dfrac{3}{x-2}$

16. $\dfrac{x+3}{y-x} = \dfrac{-(x+3)}{-(y-x)} = \dfrac{-x-3}{x-y}$

18. $-\dfrac{x-4}{x-2y} = \dfrac{x-4}{-(x-2y)} = \dfrac{x-4}{2y-x}$

20. $\dfrac{-a-1}{2b-3a} = \dfrac{-(-a-1)}{-(2b-3a)} = \dfrac{a+1}{3a-2b}$

22. $\dfrac{100mn}{-5m^2n^3} = \dfrac{5mn(20)}{-5mn(mn^2)} = \dfrac{-20}{mn^2}$

24. $\dfrac{-15xy^3z}{-3y^2z^4} = \dfrac{-3y^2z(5xy)}{-3y^2z(z^3)} = \dfrac{5xy}{z^3}$

26. $\dfrac{3a^3(2a-1)}{9a^2(2a-1)^2} = \dfrac{3a^2(2a-1)a}{3a^2(2a-1)3(2a-1)} = \dfrac{a}{3(2a-1)}$

28. $\dfrac{v^4(4v-1)}{4v(1-4v)} = \dfrac{v(1-4v)(-v^3)}{v(1-4v)(4)} = \dfrac{-v^3}{4}$

30. $\dfrac{2y-8}{8} = \dfrac{2(y-4)}{2(4)} = \dfrac{y-4}{4}$

32. $\dfrac{3x^3-6x^2+3x}{-3x} = -\dfrac{3x(x^2-2x+1)}{3x} = -(x-1)^2$

34. $\dfrac{5x^3y^5+10x^4y^2}{5x^2y^2} = \dfrac{5x^2y^2x(y^3+2x)}{5x^2y^2} = x(y^3+2x)$

36. $\dfrac{4-4x^2}{(x+1)^2} = \dfrac{4(1+x)(1-x)}{(x+1)(x+1)} = \dfrac{4(1-x)}{(x+1)}$

38. $\dfrac{5y^2-20}{2y-4} = \dfrac{5(y+2)(y-2)}{2(y-2)} = \dfrac{5(y+2)}{2}$

40 $\dfrac{4-2y}{y^3-8} = \dfrac{-2(y-2)}{(y-2)(y^2+2y+4)} = \dfrac{-2}{y^2+2y+4}$

42. $\dfrac{5x^2+10x}{5x^3+20x} = \dfrac{5x(x+2)}{5x(x^2+4)} = \dfrac{x+2}{x^2+4}$

44. $\dfrac{(2x-y)^2}{y^2-4x^2} = \dfrac{(2x-y)(2x-y)}{(y+2x)(y-2x)} = \dfrac{-(2x-y)(y-2x)}{(y+2x)(y-2x)} = \dfrac{y-2x}{y+2x}$

46. $\dfrac{6x^2-x-1}{2x^2+9x-5} = \dfrac{(3x+1)(2x-1)}{(x+5)(2x-1)} = \dfrac{3x+1}{x+5}$

48. $\dfrac{2x-30+4x^2}{15-16x+4x^2} = \dfrac{(2x-5)(2x+6)}{(2x-5)(2x-3)} = \dfrac{2x+6}{2x-3}$

50. $\dfrac{8y^3-1}{4y^2-1} = \dfrac{(2y-1)(4y^2+2y+1)}{(2y-1)(2y+1)} = \dfrac{4y^2+2y+1}{2y+1}$

52. $\dfrac{8x^2y^2 - 18xy + 7}{6x^2y^2 + 7xy - 5} = \dfrac{(4xy - 7)(2xy - 1)}{(3xy + 5)(2xy - 1)} = \dfrac{4xy - 7}{3xy + 5}$

54. $\dfrac{a^2 + 4ab + 4b^2}{2a^2 + 5ab + 2b^2} = \dfrac{(a + 2b)(a + 2b)}{(2a + b)(a + 2b)} = \dfrac{a + 2b}{2a + b}$

56. $\dfrac{6t^6 + 10t^4 - 4t^2}{6t^6 - 6t^4 - 36t^2} = \dfrac{2t^2(3t^2 - 1)(t^2 + 2)}{2t^2(3)(t^2 - 3)(t^2 + 2)} = \dfrac{3t^2 - 1}{3(t^2 - 3)}$

58. $\dfrac{12b^4c^3 - 16b^3c^2 - 3b^2c}{8b^3c^4 - 16b^2c^3 + 6bc^2} = \dfrac{b(bc)(2bc - 3)(6bc + 1)}{2c(bc)(2bc - 3)(2bc - 1)} = \dfrac{b(6bc + 1)}{2c(2bc - 1)}$

60. $\dfrac{ax^2 + x + a^2x + a}{3x + 3a} = \dfrac{x(ax + 1) + a(ax + 1)}{3(x + a)} = \dfrac{(x + a)(ax + 1)}{(x + a)(3)} = \dfrac{ax + 1}{3}$

62. $\dfrac{2ax - 4ay + bx - 2by}{x^2 - 4y^2} = \dfrac{2a(x - 2y) + b(x - 2y)}{(x + 2y)(x - 2y)} = \dfrac{(2a + b)(x - 2y)}{(x + 2y)(x - 2y)} = \dfrac{2a + b}{x + 2y}$

64. $\dfrac{3x^3 - 9x - x^2 + 3}{3x^3 + 3x - x^2 - 1} = \dfrac{x^2(3x - 1) - 3(3x - 1)}{x^2(3x - 1) + (3x - 1)} = \dfrac{(x^2 - 3)(3x - 1)}{(x^2 + 1)(3x - 1)} = \dfrac{x^2 - 3}{x^2 + 1}$

66. $\dfrac{2(x + 2)(x^2 - 1)^3 - 6x(x^2 - 1)^2(x + 2)^2}{(x^2 - 1)^6} = \dfrac{2(x + 2)(x^2 - 1)^2[(x^2 - 1) - 3x(x + 2)]}{(x^2 - 1)^6} =$

$\dfrac{2(x + 2)(-2x^2 - 6x - 1)}{(x^2 - 1)^4}$

68. $\dfrac{(2x - 3)^4(x^4 - 1) - x^3(2x - 3)(x^4 - 1)}{x^4 - x^3 + x^2 - x} = \dfrac{(2x - 3)(x^4 - 1)[(2x - 3)^3 - x^3]}{x^3(x - 1) + x(x - 1)} =$

$\dfrac{(2x - 3)(x - 1)(x + 1)(x^2 + 1)(8x^3 - 18x^2 + 54x - 27 - x^3)}{x(x^2 + 1)(x - 1)} =$

$\dfrac{(2x - 3)(x + 1)(7x^3 - 18x^2 + 54x - 27)}{x}$

70. a. No: $\dfrac{9b^2 - 3b}{3b} = \dfrac{3b(3b - 1)}{3b} = 3b - 1$ b. No: $\dfrac{b + 2}{3b^2 + 6b} = \dfrac{1(b + 2)}{3b(b + 2)} = \dfrac{1}{3b}$

c. No: $\dfrac{3b - 9}{9} = \dfrac{3(b - 3)}{3(3)} = \dfrac{b - 3}{3}$ d. Yes: $\dfrac{9b^2 - 3b}{3b - 1} = \dfrac{3b(3b - 1)}{3b - 1} = 3b$

72. a. Yes: $\dfrac{2a - b}{b - 2a} = \dfrac{-1(b - 2a)}{b - 2a} = -1$ b. Yes: $\dfrac{-b^2 - 2}{b^2 + 2} = \dfrac{-1(b^2 + 2)}{b^2 + 2} = -1$

c. No: already reduced   d. No: already reduced

74. Let x represent the speed of the current.

a. Time upstream $= \dfrac{5}{15 - x}$ b. Time downstream $= \dfrac{5}{15 + x}$

14

c. Total time = $\dfrac{5}{15 - x} + \dfrac{5}{15 + x}$

76. Let x represent the speed of the blimp.

a. time north = $\dfrac{23}{x - 6}$ b. time south = $\dfrac{23}{x + 6}$

c. The trip north takes longer by $\dfrac{23}{x - 6} - \dfrac{23}{x + 6}$ or $\dfrac{276}{(x-6)(x+6)}$ hours.

78. Let x represent the length of the panel.

a. Width = $\dfrac{200}{x}$ b. Perimeter = $2x + 2\left(\dfrac{200}{x}\right)$; no, the perimeter

depends on the length of the panel.

c. cost = $5\left(2x + 2\left(\dfrac{200}{x}\right)\right) = 10x + \dfrac{2000}{x}$

80. Let x represent the length of the freezer.

a. height = $\dfrac{30}{x(x - 2)}$

b. surface area = $2x(x - 2) + (2x)\dfrac{30}{x(x - 2)} + 2(x-2)\dfrac{30}{x(x - 2)} =$

$2x(x - 2) + \dfrac{60}{x - 2} + \dfrac{60}{x}$

Exercise 2.2

2. $\dfrac{3}{10} \cdot \dfrac{16}{27} \cdot \dfrac{30}{36} = \dfrac{(3)(4)(4)(3)(10)}{(10)(3)(3)(3)(4)(9)} = \dfrac{4}{27}$

4. $\dfrac{21t^2}{5s} \cdot \dfrac{15s^3}{7st} = \dfrac{(7)(3)tt(3)(5)sss}{5s(7)st} = 9st$

6. $\dfrac{14a^3b}{3b} \cdot \dfrac{-6}{7a^2} = \dfrac{7(2)a^2ab(-3)(2)}{3b(7)a^2} = -4a$ 8. $\dfrac{2}{3}y \cdot \dfrac{9}{10}y^2 = \dfrac{3}{5}y^3$

10. $\dfrac{1}{4}x^3y \cdot \dfrac{2}{5}xy = \dfrac{1}{10}x^4y^2$ 12. $-\dfrac{3}{5}x^2y \cdot \dfrac{5}{6}xy^2z = -\dfrac{1}{2}x^3y^3z$

14. $\dfrac{a^2}{xy} \cdot \dfrac{3x^3y}{4a} = \dfrac{3ax^2}{4}$ 16. $\dfrac{10x}{12y} \cdot \dfrac{3x^2z}{5x^3z} \cdot \dfrac{6y^2x}{3yz} = \dfrac{x}{z}$

18. $15x^2y \cdot \dfrac{3}{45xy^2} = \dfrac{x}{y}$

15

20. $\dfrac{3y}{4xy - 6y^2} \cdot \dfrac{2x - 3y}{12x} = \dfrac{3y}{2y(2x - 3y)} \cdot \dfrac{2x - 3y}{12x} = \dfrac{1}{8x}$

22. $\dfrac{9x^2 - 25}{2x - 2} \cdot \dfrac{x^2 - 1}{6x - 10} = \dfrac{(3x - 5)(3x + 5)}{2(x - 1)} \cdot \dfrac{(x - 1)(x + 1)}{2(3x - 5)} = \dfrac{(3x + 5)(x + 1)}{4}$

24. $\dfrac{3x^2 - 7x - 6}{2x^2 - x - 1} \cdot \dfrac{2x^2 - 9x - 5}{3x^2 - 13x - 10} = \dfrac{(3x + 2)(x - 3)}{(2x + 1)(x - 1)} \cdot \dfrac{(2x + 1)(x - 5)}{(3x + 2)(x - 5)} = \dfrac{x - 3}{x - 1}$

26. $\dfrac{5x^2 - 5x}{10x - 2} \cdot \dfrac{x^2 - 9x - 10}{4x - 40} \cdot \dfrac{x^2 - 2x}{2 - 2x^2} = \dfrac{5x(x - 1)}{2(5x - 1)} \cdot \dfrac{(x + 1)(x - 10)}{4(x - 10)} \cdot \dfrac{x(x - 2)}{2(1 - x)(1 + x)}$

$= -\dfrac{5x^2(x - 2)}{16(5x - 1)}$

28. $\dfrac{x^4 - 3x^3}{x^4 + 6x^2 - 27} \cdot \dfrac{x^4 - 81}{3x^4 - 81x} = \dfrac{x^3(x - 3)}{(x^2 + 9)(x^2 - 3)} \cdot \dfrac{(x^2 + 9)(x + 3)(x - 3)}{3x(x - 3)(x^2 + 3x + 9)} =$

$\dfrac{x^2(x - 3)(x + 3)}{3(x^2 - 3)(x^2 + 3x + 9)}$

30. $\dfrac{a^2 + 3ax - 3a - 9x}{3a^4 - 27a^2x^2} \cdot \dfrac{3a^4 - 11a^3x + 6a^2x^2}{3a^2x^2 - 9ax^2} = \dfrac{(a - 3)(a + 3x)}{3a^2(a - 3x)(a + 3x)} \cdot \dfrac{a^2(3a - 2x)(a - 3x)}{3x^2a(a - 3)} = \dfrac{3a - 2x}{9ax^2}$

32. $\dfrac{3}{4}y\left(\dfrac{1}{6}y + 8\right) = \left(\dfrac{3}{4}y\right)\left(\dfrac{1}{6}y\right) + \dfrac{3}{4}y(8) = \dfrac{1}{8}y^2 + 6y$
34. $\left(y - \dfrac{1}{3}\right)\left(y - \dfrac{1}{3}\right) = y^2 - \dfrac{2}{3}y + \dfrac{1}{9}$

36. $\left(y + \dfrac{1}{4}\right)^2 = y^2 + \dfrac{1}{2}y + \dfrac{1}{16}$
38. $\dfrac{6y - 27}{5x} \div \dfrac{4y - 18}{x} = \dfrac{3(2y - 9)}{5x} \cdot \dfrac{x}{2(2y - 9)} = \dfrac{3}{10}$

40. $\dfrac{a^2 + 2a - 15}{a^2 + 3a - 10} \div \dfrac{a^2 - 9}{a^2 - 9a + 14} = \dfrac{(a + 5)(a - 3)}{(a + 5)(a - 2)} \cdot \dfrac{(a - 7)(a - 2)}{(a - 3)(a + 3)} = \dfrac{a - 7}{a + 3}$

42. $\dfrac{9x^2 + 3x - 2}{12x^2 + 5x - 2} \div \dfrac{9x^2 - 6x + 1}{8x^2 + 10x - 3} = \dfrac{(3x - 1)(3x + 2)}{(4x - 1)(3x + 2)} \cdot \dfrac{(2x + 3)(4x - 1)}{(3x - 1)(3x - 1)} = \dfrac{2x + 3}{3x - 1}$

44. $\dfrac{8x^3 - y^3}{x + y} \div \dfrac{2x - y}{x^2 - y^2} = \dfrac{(2x - y)(4x^2 + 2xy + y^2)}{x + y} \cdot \dfrac{(x - y)(x + y)}{2x - y} =$

$(4x^2 + 2xy + y^2)(x - y)$

46. $\dfrac{2xy + 4x + 3y + 6}{2x^2 + x - 3} \div \dfrac{y^2 + 4y + 4}{y^2 - 3y + 2} = \dfrac{(2x + 3)(y + 2)}{(2x + 3)(x - 1)} \cdot \dfrac{(y - 2)(y - 1)}{(y + 2)(y + 2)} = \dfrac{(y - 2)(y - 1)}{(x - 1)(y + 2)}$

48. $1 \div \dfrac{x^2 + 3x + 1}{x - 2} = 1 \cdot \dfrac{x - 2}{x^2 + 3x + 1} = \dfrac{x - 2}{x^2 + 3x + 1}$

50. $(x^2 - 9) \div \dfrac{x^2 - 6x + 9}{3x} = (x + 3)(x - 3)\dfrac{3x}{(x - 3)(x - 3)} = \dfrac{3x(x + 3)}{x - 3}$

16

52. $\dfrac{2y^2 + y}{3x} \div 2y = \dfrac{y(2y + 1)}{3x} \cdot \dfrac{1}{2y} = \dfrac{2y + 1}{6x}$

54. $\dfrac{8a^2x^2 - 4ax^2 + ax}{2ax} = \dfrac{8a^2x^2}{2ax} - \dfrac{4ax^2}{2ax} + \dfrac{ax}{2ax} = 4ax - 2x + \dfrac{1}{2}$

56. $\dfrac{25m^6 - 15m^4 + 7}{-5m^3} = \dfrac{25m^6}{-5m^3} - \dfrac{15m^4}{-5m^3} + \dfrac{7}{-5m^3} = -5m^3 + 3m - \dfrac{7}{5m^3}$

58. $\dfrac{36s^4t^5 + 24s^3t^3 - s^2t}{12st^2} = \dfrac{36s^4t^5}{12st^2} + \dfrac{24s^3t^3}{12st^2} - \dfrac{s^2t}{12st^2} = 3s^3t^3 + 2s^2t - \dfrac{s}{12t}$

60. $\dfrac{4t^2 - 4t - 5}{2t - 1} = 2t - 1 - \dfrac{6}{2t - 1}$ 62. $\dfrac{2x^3 - 3x^2 - 2x + 4}{x + 1} = 2x^2 - 5x + 3 + \dfrac{1}{x + 1}$

64. $\dfrac{7 - 3t^3 - 23t^2 + 10t^4}{2t + 3} = 5t^3 - 9t^2 + 2t - 3 + \dfrac{16}{2t + 3}$

66. $\dfrac{y^5 + 1}{y - 1} = y^4 + y^3 + y^2 + y + 1 + \dfrac{2}{y - 1}$ 68. $\dfrac{2y^3 + 5y^2 - 3y + 2}{y^2 - y - 3} = 2y + 7 + \dfrac{10y + 23}{y^2 - y - 3}$

70. $\dfrac{2b^4 - 3b^2 + b + 2}{b^2 + b - 3} = 2b^2 - 2b + 5 + \dfrac{-10b + 17}{b^2 + b - 3}$

72. $\dfrac{r^4 + r^3 - 2r^2 + r + 5}{r^3 + 2r + 3} = r + 1 + \dfrac{-4r^2 - 4r + 2}{r^3 + 2r + 3}$

74. Remainder under long division is k - 9, so let k = 9.

Exercise 2.3

2. $\dfrac{y}{7} - \dfrac{5}{7} = \dfrac{y - 5}{7}$ 4. $\dfrac{1}{3}x - \dfrac{2}{3}y + \dfrac{1}{3}z = \dfrac{x}{3} - \dfrac{2y}{3} + \dfrac{z}{3} = \dfrac{x - 2y + z}{3}$

6. $\dfrac{y + 1}{b} + \dfrac{y - 1}{b} = \dfrac{2y}{b}$ 8. $\dfrac{2}{a - 3b} - \dfrac{b - 2}{a - 3b} + \dfrac{b}{a - 3b} = \dfrac{2 - (b - 2) + b}{a - 3b} = \dfrac{4}{a - 3b}$

10. $\dfrac{x + 4}{x^2 - x + 2} - \dfrac{2x - 3}{x^2 - x + 2} = \dfrac{x + 4 - (2x - 3)}{x^2 - x + 2} = \dfrac{-x + 7}{x^2 - x + 2}$ 12. $\dfrac{5}{3y} = \dfrac{7(5)}{7(3y)} = \dfrac{35}{21y}$

14. $\dfrac{-a}{b} = \dfrac{-a(ab)}{b(ab)} = \dfrac{-a^2b}{ab^2}$ 16. $x = \dfrac{x(xy^3)}{xy^3} = \dfrac{x^2y^3}{xy^3}$

18. $\dfrac{5y}{y + 3} = \dfrac{5y(y - 2)}{(y + 3)(y - 2)} = \dfrac{5y^2 - 10y}{y^2 + y - 6}$ 20. $\dfrac{5}{2a + b} = \dfrac{5(b - 2a)}{(b + 2a)(b - 2a)} = \dfrac{5b - 10a}{b^2 - 4a^2}$

22. $8(a - b)^2 = 4(2)(a - b)^2$, $12a^2b^2 = 4(3)a^2b^2$, so the LCD is $4(2)(3)(a - b)^2a^2b^2$ or

$24(a - b)^2a^2b^2$

24. $x^2 - 3x + 2 = (x - 1)(x - 2)$, $(x - 1)^2 = (x - 1)^2$, so the LCD is $(x - 1)^2(x - 2)$

1 7

26. $y^2 + 2y = y(y + 2)$, $(y + 2)^2 = (y + 2)^2$, so the LCD is $y(y + 2)^2$

28. $9y = 3(3)y$, $6y^3 - 6y = 2(3)y(y - 1)(y + 1)$, $(y - 1)^3 = (y - 1)^3$, so the LCD is

$2(3)(3)y(y - 1)^3(y + 1)$ or $18y(y - 1)^3(y + 1)$

30. $\dfrac{3y}{4} + \dfrac{y}{3} = \dfrac{9y}{12} + \dfrac{4y}{12} = \dfrac{13y}{12}$

32. $\dfrac{y}{2} + \dfrac{2y}{3} - \dfrac{3y}{4} = \dfrac{6y}{12} + \dfrac{8y}{12} - \dfrac{9y}{12} = \dfrac{5y}{12}$

34. $\dfrac{3}{4}x - \dfrac{1}{6}x = \dfrac{9}{12}x - \dfrac{2}{12}x = \dfrac{7}{12}x$

36. $\dfrac{3}{4}y + \dfrac{1}{3}y - \dfrac{5}{6}y = \dfrac{9}{12}y + \dfrac{4}{12}y - \dfrac{10}{12}y = \dfrac{3}{12}y = \dfrac{1}{4}y$

38. $\dfrac{y - 2}{4y} + \dfrac{2x - 3}{3x} = \dfrac{(y - 2)3x}{12xy} + \dfrac{4y(2x - 3)}{12xy} = \dfrac{3xy - 6x + 8xy - 12y}{12xy} = \dfrac{11xy - 6x - 12y}{12xy}$

40. $\dfrac{2}{y + 2} + \dfrac{3}{y - 2} = \dfrac{2(y - 2)}{(y + 2)(y - 2)} + \dfrac{3(y + 2)}{(y + 2)(y - 2)} = \dfrac{2y - 4 + 3y + 6}{(y + 2)(y - 2)} = \dfrac{5y + 2}{(y + 2)(y - 2)}$

42. $\dfrac{2x}{3x + 1} - \dfrac{x}{x - 2} = \dfrac{2x(x - 2)}{(3x + 1)(x - 2)} - \dfrac{x(3x + 1)}{(3x + 1)(x - 2)} = \dfrac{2x^2 - 4x - 3x^2 - x}{(3x + 1)(x - 2)} = \dfrac{-x^2 - 5x}{(3x + 1)(x - 2)}$

44. $\dfrac{x - 2}{2x + 1} - \dfrac{x + 1}{x - 1} = \dfrac{(x - 2)(x - 1)}{(2x + 1)(x - 1)} - \dfrac{(x + 1)(2x + 1)}{(2x + 1)(x - 1)} = \dfrac{x^2 - 3x + 2 - (2x^2 + 3x + 1)}{(2x + 1)(x - 1)} =$

$\dfrac{-x^2 - 6x + 1}{(2x + 1)(x - 1)}$

46. $\dfrac{2}{3y + 6} - \dfrac{3}{2y + 4} = \dfrac{2}{3(y + 2)} - \dfrac{3}{2(y + 2)} = \dfrac{4}{6(y + 2)} - \dfrac{9}{6(y + 2)} = \dfrac{-5}{6(y + 2)}$

48. $\dfrac{1}{y^2 - 1} + \dfrac{1}{y^2 + 2y + 1} = \dfrac{y + 1}{(y + 1)^2(y - 1)} + \dfrac{y - 1}{(y + 1)^2(y - 1)} = \dfrac{2y}{(y + 1)^2(y - 1)}$

50. $\dfrac{x}{x^2 - 5x + 6} - \dfrac{x - 1}{x^2 - 9} = \dfrac{x(x + 3)}{(x - 2)(x - 3)(x + 3)} - \dfrac{(x - 1)(x - 2)}{(x - 2)(x - 3)(x + 3)} =$

$\dfrac{x^2 + 3x - (x^2 - 3x + 2)}{(x - 2)(x - 3)(x + 3)} = \dfrac{6x - 2}{(x - 2)(x - 3)(x + 3)}$

52. $\dfrac{x + 1}{x^2 + 2x} - \dfrac{x - 1}{x^2 - 3x} = \dfrac{(x + 1)(x - 3)}{x(x + 2)(x - 3)} - \dfrac{(x - 1)(x + 2)}{x(x + 2)(x - 3)} = \dfrac{x^2 - 2x - 3 - (x^2 + x - 2)}{x(x + 2)(x - 3)} =$

$\dfrac{-3x - 1}{x(x + 2)(x - 3)}$

54. $\dfrac{3y - 1}{y^2 - 4y + 3} - \dfrac{y + 2}{(y - 3)^2} = \dfrac{(3y - 1)(y - 3)}{(y - 3)^2(y - 1)} - \dfrac{(y + 2)(y - 1)}{(y - 3)^2(y - 1)} = \dfrac{3y^2 - 10y + 3 - (y^2 + y - 2)}{(y - 3)^2(y - 1)} =$

$\dfrac{2y^2 - 11y + 5}{(y - 3)^2(y - 1)} = \dfrac{(2y - 1)(y - 5)}{(y - 3)^2(y - 1)}$

56. $\dfrac{4}{a^2-4}+\dfrac{2}{a^2+3a+2}+\dfrac{4}{a^2-a-2}=\dfrac{4(a+1)}{(a-2)(a+2)(a+1)}+\dfrac{2(a-2)}{(a-2)(a+2)(a+1)}+$

$\dfrac{4(a+2)}{(a-2)(a+2)(a+1)}=\dfrac{4a+4+2a-4+4a+8}{(a-2)(a+2)(a+1)}=\dfrac{10a+8}{(a-2)(a+2)(a+1)}$

58. $\dfrac{2z+5}{z^2+5z+4}+\dfrac{z+13}{z^2-z-20}+\dfrac{z+7}{z^2-4z-5}=\dfrac{(2z+5)(z-5)}{(z+4)(z+1)(z-5)}+\dfrac{(z+13)(z+1)}{(z+4)(z+1)(z-5)}+$

$\dfrac{(z+7)(z+4)}{(z+4)(z+1)(z-5)}=\dfrac{2z^2-5z-25+z^2+14z+13+z^2+11z+28}{(z+4)(z+1)(z-5)}=$

$\dfrac{4z^2+20z+16}{(z+4)(z+1)(z-5)}=\dfrac{4(z+4)(z+1)}{(z+4)(z+1)(z-5)}=\dfrac{4}{z-5}$

60. $1+\dfrac{1}{y}=\dfrac{y}{y}+\dfrac{1}{y}=\dfrac{y+1}{y}$

62. $y-\dfrac{2}{y^2-1}+\dfrac{3}{y+1}=\dfrac{y}{1}-\dfrac{2}{y^2-1}+\dfrac{3}{y+1}=\dfrac{y(y-1)(y+1)}{(y-1)(y+1)}-\dfrac{2}{(y-1)(y+1)}+\dfrac{3(y-1)}{(y-1)(y+1)}=$

$\dfrac{y^3-y-2+3y-3}{(y-1)(y+1)}=\dfrac{y^3+2y-5}{(y-1)(y+1)}$

64. $x+\dfrac{2x^2}{x+2}-\dfrac{3x^2}{x-1}=\dfrac{x}{1}+\dfrac{2x^2}{x+2}-\dfrac{3x^2}{x-1}=\dfrac{x(x+2)(x-1)+2x^2(x-1)-3x^2(x+2)}{(x+2)(x-1)}=$

$\dfrac{x^3+x^2-2x+2x^3-2x^2-3x^3-6x^2}{(x+2)(x-1)}=\dfrac{-7x^2-2x}{(x+2)(x-1)}$

66. $x+3+\dfrac{1}{x-1}=\dfrac{x+3}{1}+\dfrac{1}{x-1}=\dfrac{(x+3)(x-1)+1}{x-1}=\dfrac{x^2+2x-3+1}{x-1}=\dfrac{x^2+2x-2}{x-1}$

68. $\left(\dfrac{x}{x^2+1}-\dfrac{1}{x+1}\right)\cdot\dfrac{x+1}{x-1}=\dfrac{x(x+1)-(x^2+1)}{(x^2+1)(x+1)}\cdot\dfrac{x+1}{x-1}=\dfrac{(x^2+x-x^2-1)(x+1)}{(x^2+1)(x+1)(x-1)}=$

$\dfrac{x-1}{(x^2+1)(x-1)}=\dfrac{1}{x^2+1}$

70. $\left(\dfrac{x}{x+1}+\dfrac{1}{x-1}\right)\div\dfrac{x^2+1}{x+1}=\dfrac{x(x-1)+1(x+1)}{(x+1)(x-1)}\cdot\dfrac{x+1}{x^2+1}=\dfrac{(x^2-x+x+1)(x+1)}{(x+1)(x-1)(x^2+1)}=$

$\dfrac{1(x^2+1)}{(x-1)(x^2+1)}=\dfrac{1}{x-1}$

Exercise 2.4

2. $\dfrac{\dfrac{5}{2}}{\dfrac{21}{4}} = \dfrac{5}{2} \div \dfrac{21}{4} = \dfrac{5}{2} \cdot \dfrac{4}{21} = \dfrac{10}{21}$

4. $\dfrac{\dfrac{3ab}{4}}{\dfrac{3b}{8a^2}} = \dfrac{3ab}{4} \div \dfrac{3b}{8a^2} = \dfrac{3ab}{4} \cdot \dfrac{8a^2}{3b} = 2a^3$

6. $\dfrac{\dfrac{1}{3}}{4 + \dfrac{2}{3}} = \dfrac{3\left(\dfrac{1}{3}\right)}{3\left(4 + \dfrac{2}{3}\right)} = \dfrac{1}{3(4) + 3\left(\dfrac{2}{3}\right)} = \dfrac{1}{12 + 2} = \dfrac{1}{14}$

8. $\dfrac{\dfrac{1}{2} + \dfrac{3}{4}}{\dfrac{1}{2} - \dfrac{3}{4}} = \dfrac{4\left(\dfrac{1}{2} + \dfrac{3}{4}\right)}{4\left(\dfrac{1}{2} - \dfrac{3}{4}\right)} = \dfrac{4\left(\dfrac{1}{2}\right) + 4\left(\dfrac{3}{4}\right)}{4\left(\dfrac{1}{2}\right) - 4\left(\dfrac{3}{4}\right)} = \dfrac{2 + 3}{2 - 3} = \dfrac{5}{-1} = -5$

10. $\dfrac{\dfrac{2}{y} + \dfrac{1}{2y}}{y + \dfrac{y}{2}} = \dfrac{2y\left(\dfrac{2}{y} + \dfrac{1}{2y}\right)}{2y\left(y + \dfrac{y}{2}\right)} = \dfrac{2y\left(\dfrac{1}{y}\right) + 2y\left(\dfrac{1}{2y}\right)}{2y(y) + 2y\left(\dfrac{y}{2}\right)} = \dfrac{4 + 1}{2y^2 + y^2} = \dfrac{5}{3y^2}$

12. $\dfrac{4 - \dfrac{1}{x^2}}{2 - \dfrac{1}{x}} = \dfrac{x^2\left(4 - \dfrac{1}{x^2}\right)}{x^2\left(2 - \dfrac{1}{x}\right)} = \dfrac{4x^2 - 1}{2x^2 - x} = \dfrac{(2x - 1)(2x + 1)}{x(2x - 1)} = \dfrac{2x + 1}{x}$

14. $\dfrac{1 + \dfrac{1}{x}}{1 - \dfrac{1}{x}} = \dfrac{x\left(1 + \dfrac{1}{x}\right)}{x\left(1 - \dfrac{1}{x}\right)} = \dfrac{x + 1}{x - 1}$

16. $\dfrac{4}{\dfrac{2}{x} + 2} = \dfrac{4x}{x\left(\dfrac{2}{x} + 2\right)} = \dfrac{2(2x)}{2(x + 1)} = \dfrac{2x}{x + 1}$

18. $\dfrac{y + 3}{\dfrac{9}{y} - y} = \dfrac{y(y + 3)}{y\left(\dfrac{9}{y} - y\right)} = \dfrac{y(y + 3)}{9 - y^2} = \dfrac{y(y + 3)}{(3 - y)(3 + y)} = \dfrac{y}{3 - y}$

20. $\dfrac{x - y}{\dfrac{x}{y} - \dfrac{y}{x}} = \dfrac{xy(x - y)}{xy\left(\dfrac{x}{y} - \dfrac{y}{x}\right)} = \dfrac{xy(x - y)}{x^2 - y^2} = \dfrac{xy(x - y)}{(x - y)(x + y)} = \dfrac{xy}{x + y}$

22. $\dfrac{y + \dfrac{x}{y}}{x - \dfrac{y}{x}} = \dfrac{xy\left(y + \dfrac{x}{y}\right)}{xy\left(x - \dfrac{y}{x}\right)} = \dfrac{xy^2 + x^2}{x^2y - y^2} = \dfrac{x(x + y^2)}{y(x^2 - y)}$

24. $\dfrac{\dfrac{6}{b} - \dfrac{6}{a}}{\dfrac{3}{a^2} - \dfrac{3}{b^2}} = \dfrac{a^2b^2\left(\dfrac{6}{b} - \dfrac{6}{a}\right)}{a^2b^2\left(\dfrac{3}{a^2} - \dfrac{3}{b^2}\right)} = \dfrac{6a^2b - 6ab^2}{3b^2 - 3a^2} = \dfrac{6ab(a-b)}{3(b-a)(b+a)} = \dfrac{-2ab}{a+b}$

26. $\dfrac{1 - \dfrac{2y}{x}}{x - \dfrac{4y^2}{x}} = \dfrac{x\left(1 - \dfrac{2y}{x}\right)}{x\left(x - \dfrac{4y^2}{x}\right)} = \dfrac{x - 2y}{x^2 - 4y^2} = \dfrac{1(x-2y)}{(x-2y)(x+2y)} = \dfrac{1}{x+2y}$

28. $\dfrac{\dfrac{4}{b^2c} - \dfrac{5}{bc^2}}{\dfrac{10}{b} - \dfrac{8}{c}} = \dfrac{b^2c^2\left(\dfrac{4}{b^2c} - \dfrac{5}{bc^2}\right)}{b^2c^2\left(\dfrac{10}{b} - \dfrac{8}{c}\right)} = \dfrac{4c - 5b}{10bc^2 - 8b^2c} = \dfrac{4c - 5b}{2bc(5c - 4b)}$

30. $\dfrac{6 + \dfrac{1}{z} - \dfrac{2}{z^2}}{9 + \dfrac{6}{z}} = \dfrac{z^2\left(6 + \dfrac{1}{z} - \dfrac{2}{z^2}\right)}{z^2\left(9 + \dfrac{6}{z}\right)} = \dfrac{6z^2 + z - 2}{9z^2 + 6z} = \dfrac{(3z+2)(2z-1)}{3z(3z+2)} = \dfrac{2z-1}{3z}$

32. $\dfrac{\dfrac{1}{y-1}}{\dfrac{1}{y^2} + 1} = \dfrac{y^2(y-1)\dfrac{1}{y-1}}{y^2(y-1)\left(\dfrac{1}{y^2} + 1\right)} = \dfrac{y^2}{(y-1)(1+y^2)}$

34. $\dfrac{c + 1 - \dfrac{10}{c-2}}{\dfrac{-2}{c-2} + c - 3} = \dfrac{(c-2)\left(c + 1 - \dfrac{10}{c-2}\right)}{(c-2)\left(\dfrac{-2}{c-2} + c - 3\right)} = \dfrac{c^2 - c - 2 - 10}{c^2 - 5c + 6 - 2} = \dfrac{(c-4)(c+3)}{(c-4)(c-1)} = \dfrac{c+3}{c-1}$

36. $\dfrac{\dfrac{3}{v-2} - \dfrac{1}{v+2}}{\dfrac{1}{v-2} - \dfrac{3}{v-3}} = \dfrac{(v-2)(v+2)(v-3)\left(\dfrac{3}{v-2} - \dfrac{1}{v+2}\right)}{(v-2)(v+2)(v-3)\left(\dfrac{1}{v-2} - \dfrac{3}{v-3}\right)} = \dfrac{3v^2 - 3v - 18 - (v^2 - 5v + 6)}{v^2 - v - 6 - (3v^2 - 12)} =$

$\dfrac{2v^2 + 2v - 24}{-2v^2 - v + 6} = \dfrac{2(v+4)(v-3)}{(v+2)(-2v+3)}$

38. $\dfrac{\dfrac{1}{z-1} - \dfrac{z}{z+1}}{1 - \dfrac{z}{z+1}} = \dfrac{(z-1)(z+1)\left(\dfrac{1}{z-1} - \dfrac{z}{z+1}\right)}{(z-1)(z+1)\left(1 - \dfrac{z}{z+1}\right)} = \dfrac{z+1 - z(z-1)}{z^2 - 1 - z(z-1)} = \dfrac{-z^2 + 2z + 1}{z-1}$

40. $\dfrac{\dfrac{x}{x-y} - \dfrac{y}{x+y}}{x^2 - y^2} = \dfrac{(x-y)(x+y)\left(\dfrac{x}{x-y} - \dfrac{y}{x+y}\right)}{(x-y)(x+y)(x^2 - y^2)} = \dfrac{x(x+y) - y(x-y)}{(x-y)(x+y)(x-y)(x+y)} =$

$\dfrac{x^2 + xy - xy + y^2}{(x-y)^2(x+y)^2} = \dfrac{x^2 + y^2}{(x-y)^2(x+y)^2}$

42. $\left(\dfrac{4}{x} - \dfrac{1}{3}\right) \div \left(\dfrac{16}{x^2} - \dfrac{1}{9}\right) = \dfrac{\dfrac{4}{x} - \dfrac{1}{3}}{\dfrac{16}{x^2} - \dfrac{1}{9}} = \dfrac{9x^2\left(\dfrac{4}{x} - \dfrac{1}{3}\right)}{9x^2\left(\dfrac{16}{x^2} - \dfrac{1}{9}\right)} = \dfrac{3x(12 - x)}{(12 - x)(12 + x)} = \dfrac{3x}{12 + x}$

44. $\left(1 + \dfrac{1}{b^3}\right) \div \left(1 + \dfrac{1}{b}\right) = \dfrac{1 + \dfrac{1}{b^3}}{1 + \dfrac{1}{b}} = \dfrac{b^3\left(1 + \dfrac{1}{b^3}\right)}{b^3\left(1 + \dfrac{1}{b}\right)} = \dfrac{b^3 + 1}{b^3 + b^2} = \dfrac{(b + 1)(b^2 - b + 1)}{b^2(b + 1)} = \dfrac{b^2 - b + 1}{b^2}$

46. $\left(y - 1 - \dfrac{4}{3y - 2}\right) \div \left(y - \dfrac{1}{3y - 2}\right) = \dfrac{y - 1 - \dfrac{4}{3y - 2}}{y - \dfrac{1}{3y - 2}} = \dfrac{(3y - 2)\left(y - 1 - \dfrac{4}{3y - 2}\right)}{(3y - 2)\left(y - \dfrac{1}{3y - 2}\right)} = \dfrac{3y^2 - 5y + 2 - 4}{3y^2 - 2y - 1} =$

$\dfrac{(3y + 1)(y - 2)}{(3y + 1)(y - 1)} = \dfrac{y - 2}{y - 1}$

48. a. $\dfrac{1}{R} = \dfrac{1}{R_1} + \dfrac{1}{R_1 + 10}$ $\qquad$ b. $R = \dfrac{1}{\dfrac{1}{R_1} + \dfrac{1}{R_1 + 10}} =$

$\dfrac{R_1(R_1 + 10)}{R_1(R_1 + 10)\left(\dfrac{1}{R_1} + \dfrac{1}{R_1 + 10}\right)} = \dfrac{R_1(R_1 + 10)}{R_1 + 10 + R_1} = \dfrac{R_1(R_1 + 10)}{2R_1 + 10}$

50. a. large press: $\dfrac{1}{t_1}$; $\quad$ small press: $\dfrac{1}{t_2}$ $\qquad$ b. $\dfrac{1}{t_1} + \dfrac{1}{t_2}$

c. $\dfrac{1}{\dfrac{1}{4} + \dfrac{1}{6}}$ $\qquad$ d. $\dfrac{1}{\dfrac{1}{4} + \dfrac{1}{6}} = \dfrac{12}{12\left(\dfrac{1}{4} + \dfrac{1}{6}\right)} = \dfrac{12}{3 + 2} = 2.4$ hours

52. $\dfrac{1 - \dfrac{1}{\dfrac{a}{b} + 2}}{1 + \dfrac{3}{\dfrac{a}{2b} + 1}} = \dfrac{1 - \dfrac{b}{b\left(\dfrac{a}{b} + 2\right)}}{1 + \dfrac{3(2b)}{2b\left(\dfrac{a}{2b} + 1\right)}} = \dfrac{1 - \dfrac{b}{a + 2b}}{1 + \dfrac{6b}{a + 2b}} = \dfrac{(a + 2b)\left(1 - \dfrac{b}{a + 2b}\right)}{(a + 2b)\left(1 + \dfrac{6b}{a + 2b}\right)} = \dfrac{a + 2b - b}{a + 2b + 6b} =$

$\dfrac{a + b}{a + 8b}$

54. $1 - \dfrac{1 - \dfrac{1}{x}}{x - \dfrac{1}{x}} = 1 - \dfrac{x\left(1 - \dfrac{1}{x}\right)}{x\left(x - \dfrac{1}{x}\right)} = 1 - \dfrac{x - 1}{x^2 - 1} = 1 - \dfrac{x - 1}{(x - 1)(x + 1)} = \dfrac{x + 1}{x + 1} - \dfrac{1}{x + 1} = \dfrac{x}{x + 1}$

56.
$$\dfrac{\dfrac{a}{bc} - \dfrac{b}{ac} + \dfrac{c}{ab}}{\dfrac{1}{a^2b^2} - \dfrac{1}{a^2c^2} + \dfrac{1}{b^2c^2}} = \dfrac{a^2b^2c^2\left(\dfrac{a}{bc} - \dfrac{b}{ac} + \dfrac{c}{ab}\right)}{a^2b^2c^2\left(\dfrac{1}{a^2b^2} - \dfrac{1}{a^2c^2} + \dfrac{1}{b^2c^2}\right)} = \dfrac{a^3bc - ab^3c + abc^3}{c^2 - b^2 + a^2} =$$

$$\dfrac{abc(a^2 - b^2 + c^2)}{a^2 - b^2 + c^2} = abc$$

Exercise 2.5

2. $\dfrac{y^2}{y^6} = \dfrac{1}{y^{6-2}} = \dfrac{1}{y^4}$ **4.** $\dfrac{x^4y^6}{x^2y} = x^{4-2}y^{6-1} = x^2y^5$ **6.** $\left(\dfrac{y^2}{z^3}\right)^2 = \dfrac{y^4}{z^6}$

8. $\left(\dfrac{-x^2}{2y}\right)^4 = \dfrac{(-1)^4(x^2)^4}{2^4y^4} = \dfrac{x^8}{16y^4}$ **10.** $\dfrac{(5x)^2}{(-3x^2)^3} = \dfrac{5^2x^2}{(-3)^3(x^2)^3} = -\dfrac{25}{27x^4}$

12. $\dfrac{(-x)^2(-x^2)^4}{(x^2)^3} = \dfrac{x^2x^8}{x^6} = x^{2+8-6} = x^4$

14. $\left(\dfrac{x^2z}{2}\right)^2\left(\dfrac{-2}{x^2z}\right)^3 = \left(\dfrac{x^4z^2}{4}\right)\left(\dfrac{-8}{x^6z^3}\right) = \dfrac{-2}{x^{6-4}z^{3-2}} = \dfrac{-2}{x^2z}$

16. $\left(\dfrac{2x-y}{y}\right)^3\left(\dfrac{-3}{2x-y}\right)^3 = \dfrac{(2x-y)^3}{y^3} \cdot \dfrac{-27}{(2x-y)^3} = \dfrac{-27}{y^3}$

18. $3^{-2} = \dfrac{1}{3^2} = \dfrac{1}{9}$ **20.** $\dfrac{3}{4^{-2}} = 3(4^2) = 3(16) = 48$ **22.** $(-5)^{-2} = \dfrac{1}{(-5)^2} = \dfrac{1}{25}$

24. $\dfrac{1}{(-3)^{-3}} = (-3)^3 = -27$ **26.** $\left(\dfrac{1}{3}\right)^{-2} = 3^2 = 9$ **28.** $\dfrac{3^{-3}}{6^{-2}} = \dfrac{6^2}{3^3} = \dfrac{36}{27} = \dfrac{4}{3}$

30. $5^{-1} + 25^0 = \dfrac{1}{5} + 1 = \dfrac{1}{5} + \dfrac{5}{5} = \dfrac{6}{5}$ **32.** $8^{-2} - 2^0 = \dfrac{1}{8^2} - 1 = \dfrac{1}{64} - \dfrac{64}{64} = -\dfrac{63}{64}$

34. $-3x^{-2} = \dfrac{-3}{x^2}$ **36.** $\dfrac{-6y^{-3}}{x^{-3}} = \dfrac{-6x^3}{y^3}$ **38.** $(x + y)^{-3} = \dfrac{1}{(x + y)^3}$

40. $\dfrac{(a-3)^{-2}}{(a+3)^{-3}} = \dfrac{(a+3)^3}{(a-3)^2}$ **42.** $x^{-5}x^2 = x^{-5+2} = x^{-3} = \dfrac{1}{x^3}$

44. $\dfrac{x^5}{x^{-5}} = x^{5-(-5)} = x^{10}$ **46.** $(2x^3y^{-4})^{-3} = 2^{-3}x^{-9}y^{12} = \dfrac{y^{12}}{8x^9}$

48. $\dfrac{x^{-3}y^{-2}}{6x^{-5}y^0} = \dfrac{x^5}{6x^3y^2y^0} = \dfrac{x^2}{6y^2}$ **50.** $\left(\dfrac{a^4}{b^{-5}}\right)^{-3} = \dfrac{a^{-12}}{b^{15}} = \dfrac{1}{a^{12}b^{15}}$

52. $\dfrac{(a-b)^{-5}}{(a-b)^{-7}} = (a-b)^{-5-(-7)} = (a-b)^2$ **54.** $y^{-2}(x^2 + y) = x^2y^{-2} + y^{-1}$

56. $x^{-2}y^{-3}(x^3 + y^3) = x^{-2}x^3y^3 + x^{-2}y^{-3}y^3 = xy^{-3} + x^{-2}$

58. $xy^{-1} + x^{-1}y^{-2} = x^{-1}y^{-2}(x^2y + 1)$ 60. $4y^{-1} + x^2y^{-3} = y^{-3}(4y^2 + x^2)$

62. $x^{-1} - y^{-3} = \dfrac{1}{x} - \dfrac{1}{y^3} = \dfrac{y^3}{xy^3} - \dfrac{x}{xy^3} = \dfrac{y^3 - x}{xy^3}$ 64. $\dfrac{x^{-1}}{y^{-1}} + \dfrac{y}{x} = \dfrac{y}{x} + \dfrac{y}{x} = \dfrac{2y}{x}$

66. $xy^{-1} + x^{-1}y = \dfrac{x}{y} + \dfrac{y}{x} = \dfrac{x^2}{xy} + \dfrac{y^2}{xy} = \dfrac{x^2 + y^2}{xy}$

68. $\dfrac{x + y^{-1}}{y^{-1}} = \dfrac{x + \dfrac{1}{y}}{\dfrac{1}{y}} = \dfrac{y\left(x + \dfrac{1}{y}\right)}{y\left(\dfrac{1}{y}\right)} = xy + 1$

70. $\dfrac{x^{-2} - y^{-2}}{(xy)^{-1}} = \left(\dfrac{1}{x^2} - \dfrac{1}{y^2}\right)(xy) = \dfrac{xy}{x^2} - \dfrac{xy}{y^2} = \dfrac{y}{x} - \dfrac{x}{y} = \dfrac{y^2}{xy} - \dfrac{x^2}{xy} = \dfrac{y^2 - x^2}{xy}$

72. $\dfrac{(x + y)^{-1}}{x^{-1} + y^{-1}} = \dfrac{1}{(x + y)\left(\dfrac{1}{x} + \dfrac{1}{y}\right)} = \dfrac{xy}{xy(x + y)\left(\dfrac{1}{x} + \dfrac{1}{y}\right)} = \dfrac{xy}{(x + y)(y + x)} = \dfrac{xy}{(x + y)^2}$

74. $68{,}742 = 6.8742 \times 10^4$ 76. $481{,}000 = 4.81 \times 10^5$

78. $0.421 = 4.21 \times 10^{-1}$ 80. $0.000004 = 4 \times 10^{-6}$ 82. $4.8 \times 10^3 = 4{,}800$

84. $8.31 \times 10^4 = 83{,}100$ 86. $8.0 \times 10^{-1} = 0.8$ 88. $4.31 \times 10^{-5} = 0.0000431$

90.   a. $\dfrac{0.0054 \times 0.05 \times 300}{0.0016 \times 0.27 \times 8200} = \dfrac{5 \times 10^{-3} \times 5 \times 10^{-2} \times 3 \times 10^2}{1.6 \times 10^{-3} \times 2.7 \times 10^{-1} \times 8.2 \times 10^3} \approx$

               $\dfrac{2 \times 5 \times 3}{1.5 \times 8 \times 10^2} = 2.5 \times 10^{-2}$

         b. $\dfrac{0.0054 \times 0.05 \times 300}{0.0016 \times 0.27 \times 8200} = 0.022866 = 2.29 \times 10^{-2}$

92.   a. $\dfrac{0.004 \times 27{,}000 \times 620{,}000}{2700 \times 0.0001 \times 0.009} = \dfrac{4 \times 10^{-3} \times 2.7 \times 10^4 \times 6.2 \times 10^5}{2.7 \times 10^3 \times 1 \times 10^{-4} \times 9 \times 10^{-3}} \approx$

               $\dfrac{4 \times 6 \times 10^6}{10 \times 10^{-4}} = 2.4 \times 10^{10}$

         b. $\dfrac{0.004 \times 27{,}000 \times 620{,}000}{2700 \times 0.0001 \times 0.009} = 2.76 \times 10^{10}$

94.　a.　1.823103×10^{12}

　　b.　$\dfrac{1.823103 \times 10^{12}}{2.38631 \times 10^8} = 7.64 \times 10^3 = \$7{,}640.00$

96.　a.　3.17×10^4 sq. mi. $= (3.17 \times 10^4)(5.28 \times 10^3)^2$ sq. ft. $= 8.84 \times 10^{11}$

　　　sq. ft.;　volume $= (8.8375 \times 10^{11})(4.83 \times 10^2) = 4.27 \times 10^{14}$ cu. ft.

　　b.　$(4.2685 \times 10^{14})(7.48) = 3.19 \times 10^{15}$ gallons

98.　a.　mass $= 6.5856 \times 10^{21}$ tons, volume $= 2.598753 \times 10^{11}$ cu. mi.

　　b.　density $= \dfrac{6.5856 \times 10^{21}}{2.598753 \times 10^{11}} = 2.53 \times 10^{10}$ tons per cu. mi.

　　c.　density $= \dfrac{2.5342 \times 10^{10} \times 2 \times 10^3}{(5.28 \times 10^3)^3} = 3.44 \times 10^2$ lb per cu. ft.

100.　$\left[\left(\dfrac{a^3bc}{x^2y}\right)^4\left(\dfrac{x^2yz}{ab^2c^3}\right)^2\right]^2 = \left[\left(\dfrac{a^{12}b^4c^4}{x^8y^4}\right)\left(\dfrac{x^4y^2z^2}{a^2b^4c^6}\right)\right]^2 = \left(\dfrac{a^{10}z^2}{x^4y^2c^2}\right)^2 = \dfrac{a^{20}z^4}{x^8y^4c^4}$

102.　$\left(\dfrac{m^3n^2p}{r^2s}\right)^2\left(\dfrac{rs}{mn^2p^2}\right)^3\left(-\dfrac{mnp}{rs}\right)^2 = \left(\dfrac{m^6n^4p^2}{r^4s^2}\right)\left(\dfrac{r^3s^3}{m^3n^6p^6}\right)\left(\dfrac{m^2n^2p^2}{r^2s^2}\right) = \dfrac{m^5}{p^2r^3s}$

104.　$\left(\dfrac{2x^{-3}z^0}{5x^{-4}z^{-2}}\right)^{-3} = \dfrac{2^{-3}x^9}{5^{-3}x^{12}z^6} = \dfrac{125}{8x^3z^6}$　　106.　$\dfrac{(3y^3z^{-2})^{-1}}{(2^{-3}y^{-2}z)^{-2}} = \dfrac{3^{-1}y^{-3}z^2}{2^6y^4z^{-2}} = \dfrac{z^4}{192y^7}$

108.　$\left(\dfrac{2y^{-3}x}{3z^2}\right)^{-2}\left(\dfrac{2x^4}{9y^{-2}z^{-2}}\right)^{-1} = \left(\dfrac{2^{-2}y^6x^{-2}}{3^{-2}z^{-4}}\right)\left(\dfrac{2^{-1}x^{-4}}{9^{-1}y^2z^2}\right) = \dfrac{(9)9z^4y^6}{(4)2x^6y^2z^2} = \dfrac{81y^4z^2}{8x^6}$

110.　$3(x-1)^{-3} - 3(3x+4)(x-1)^{-4} = 3(x-1)^{-4}[(x-1) - (3x+4)] = 3(x-1)^{-4}(-2x-5)$

112.　$3(x+7)^{-4}(2x-3)^{-3} - 6(2x-3)^{-4}(x+7)^{-3} =$

　　　$-3(x+7)^{-4}(2x-3)^{-4}[-(2x-3) + 2(x+7)] = -3(x+7)^{-4}(2x-3)^{-4}(17)$

114.　$\left(\dfrac{a}{b}\right)^n = \dfrac{a}{b} \cdot \dfrac{a}{b} \cdot \cdots \cdot \dfrac{a}{b} = \dfrac{aa\ldots a}{bb\ldots b} = \dfrac{a^n}{b^n}$

CHAPTER 3

Exercise 3.1

2. $-\sqrt{169} = -13$ 4. $\sqrt[3]{64} = 4$ 6. $-\sqrt[4]{81} = -3$

8. $\sqrt[5]{-\dfrac{100{,}000}{32}} = -\dfrac{10}{2} = -5$ 10. $\sqrt[4]{-\dfrac{1296}{625}}$ is not a real number

12. $-\sqrt[3]{\dfrac{343}{125}} = -\dfrac{7}{5}$ 14. $25^{1/2} = 5$ 16. $-81^{1/4} = -3$

18. $(-64)^{1/6}$ is not a real number   20. $(-27)^{1/3} = -3$

22. $(-243)^{-1/5} = \dfrac{1}{(-243)^{1/5}} = -\dfrac{1}{3}$ 24. $\left(\dfrac{49}{81}\right)^{1/2} = \dfrac{7}{9}$ 26. $7^{1/2} = \sqrt{7}$

28. $3x^{1/4} = 3\sqrt[4]{x}$ 30. $(3x)^{1/4} = \sqrt[4]{3x}$ 32. $6^{-1/3} = \dfrac{1}{6^{1/3}} = \dfrac{1}{\sqrt[3]{6}}$

34. $y(5x)^{-1/2} = \dfrac{y}{(5x)^{1/2}} = \dfrac{y}{\sqrt{5x}}$ 36. $(y+2)^{1/3} = \sqrt[3]{y+2}$

38. $\sqrt{5} = 5^{1/2}$ 40. $\sqrt[3]{4y} = (4y)^{1/3}$ 42. $\dfrac{2}{\sqrt[5]{3}} = \dfrac{2}{3^{1/5}} = 2(3^{-1/5})$

44. $\dfrac{2\sqrt[5]{z}}{\sqrt[3]{xy}} = \dfrac{2z^{1/5}}{(xy)^{1/3}} = 2z^{1/5}(xy)^{-1/3}$ 46. $\sqrt{x} - 2\sqrt[3]{y} = x^{1/2} - 2y^{1/3}$

48. $\dfrac{1}{\sqrt[4]{3x+2y}} = \dfrac{1}{(3x+2y)^{1/4}} = (3x+2y)^{-1/4}$

50. $\dfrac{7}{16} = 0.4375$, a terminating decimal.

52. $\dfrac{5}{12} = 0.41666...$, where the digit 6 is repeated endlessly.

54. $\dfrac{11}{13} = 0.846153846153...$, where the pattern 846153 is repeated endlessly.

56. $\dfrac{25}{6} = 4.1666...$, where the digit 6 is repeated endlessly.

26

58. $\sqrt{3} \approx 1.732$

60. $\sqrt[4]{60} \approx 2.783$

62. $\sqrt[5]{-87} \approx -2.443$

64. $\sqrt[3]{1.4} \approx 1.119$

66. $\left(\sqrt[4]{16}\right)^4 = 2^4 = 16$

68. $\left(\sqrt[3]{6}\right)^3 = 6$

70. $\left(-\sqrt[4]{7}\right)^4 = 7$

72. $\left(-\sqrt[5]{y}\right)^5 = -y$

74. $19 < \sqrt{380} < 20$, since $\sqrt{361} < \sqrt{380} < \sqrt{400}$

76. $6 < \sqrt[3]{217} < 7$, since $\sqrt[3]{216} < \sqrt[3]{217} < \sqrt[3]{343}$

78. $-4 < \sqrt[3]{-52.3} < -3$, since $\sqrt[3]{-64} < \sqrt[3]{-52.3} < \sqrt[3]{-27}$

80. $4 < \sqrt[4]{306} < 5$, since $\sqrt[4]{256} < \sqrt[4]{306} < \sqrt[4]{625}$

82. a. $225\pi = \pi r^2$, so $r = \sqrt{225} = 15$ meters

    b. $18\pi = \pi r^2$, so $r = \sqrt{18}$ inches   c. $3\pi = \pi r^2$, so $r = \sqrt{3}$ yards

    d. $0.5476\pi = \pi r^2$, so $r = \sqrt{0.5476} = 0.74$ feet

84. a. $1331\pi = \pi a^3$, so $a = \sqrt[3]{1331} = 11$ feet

    b. $260\pi = \pi a^3$, so $a = \sqrt[3]{260}$ meters   c. $23\pi = \pi a^3$, so $a = \sqrt[3]{23}$ inches

    d. $3.375\pi = \pi a^3$, so $a = \sqrt[3]{3.375} = 1.5$ centimeters

86. speed $= \sqrt{30(0.8)(160)} = \sqrt{3840} \approx 61.97$ miles per hour

88. velocity $= \sqrt[3]{\dfrac{500}{0.015}} = \sqrt[3]{\dfrac{100,000}{3}} \approx 32.183$ miles per hour

90. a. velocity $= 3960\sqrt{\dfrac{7.89 \times 10^4}{2.23 \times 10^4 + 3960}} = 3960\sqrt{\dfrac{7.89 \times 10^4}{2.626 \times 10^4}} \approx$

    $3960\sqrt{3.00457} \approx 6,864$ mi/hr

    b. velocity $= 3960\sqrt{\dfrac{7.89 \times 10^4}{2.34 \times 10^5 + 3960}} = 3960\sqrt{\dfrac{7.89 \times 10^4}{2.3796 \times 10^5}} \approx$

    $3960\sqrt{0.331568} \approx 2,280$ mi/hr

27

92. $\text{temperature} = \sqrt[4]{\dfrac{3.9 \times 10^{33}}{4\pi(6.96 \times 10^{10})^2(5.7 \times 10^{-5})}} = \sqrt[4]{\dfrac{0.00353 \times 10^{18}}{\pi}} \approx$

5.79×10^3 degrees Kelvin.

94. $\left(\dfrac{1}{16}\right)^{1/4} = \dfrac{1}{2}, \left(\dfrac{1}{16}\right)^{1/2} = \dfrac{1}{4}$, so $\left(\dfrac{1}{16}\right)^{1/4} > \left(\dfrac{1}{16}\right)^{1/2}$. In general, when n > m and

a < 1, $a^{1/n} > a^{1/m}$.

96. $\left(\dfrac{1}{2}\right)^{1/2} \approx 0.7071, \left(\dfrac{1}{2}\right)^{1/3} \approx 0.7937, \left(\dfrac{1}{2}\right)^{1/4} \approx 0.8409, \ldots, \left(\dfrac{1}{2}\right)^{1/10} \approx 0.9330,$

$\left(\dfrac{1}{2}\right)^{1/100} \approx 0.9331, \left(\dfrac{1}{2}\right)^{1/1000} \approx 0.99931; \left(\dfrac{1}{2}\right)^{1/n}$ is approaching 1.

98. $\sqrt[6]{x} = x^{1/6} = x^{(1/3)(1/2)} = \left(x^{1/3}\right)^{1/2} = \left(\sqrt[3]{x}\right)^{1/2} = \sqrt{\sqrt[3]{x}}$

Exercise 3.2

2. $125^{2/3} = \left(\sqrt[3]{125}\right)^2 = 5^2 = 25$ 4. $(-64)^{2/3} = \left(\sqrt[3]{-64}\right)^2 = (-4)^2 = 16$

6. $8^{-1/3} = \dfrac{1}{\sqrt[3]{8}} = \dfrac{1}{2}$ 8. $(-32)^{-3/5} = \dfrac{1}{\left(\sqrt[5]{-32}\right)^3} = \dfrac{1}{(-2)^3} = -\dfrac{1}{8}$

10. $\left(-\dfrac{243}{32}\right)^{2/5} = \left(\sqrt[5]{-\dfrac{243}{32}}\right)^2 = \left(-\dfrac{3}{2}\right)^2 = \dfrac{9}{4}$

12. $-\left(\dfrac{289}{100}\right)^{5/2} = -\left(\sqrt{\dfrac{289}{100}}\right)^5 = -\left(\dfrac{17}{10}\right)^5 = -\dfrac{1419857}{100000}$ or -14.19857

14. $y^{3/4} = \sqrt[4]{y^3}$ or $\left(\sqrt[4]{y}\right)^3$ 16. $5y^{2/3} = 5\sqrt[3]{y^2}$ or $5\left(\sqrt[3]{y}\right)^2$

18. $x^{-2/7} = \dfrac{1}{\sqrt[7]{x^2}}$ or $\dfrac{1}{\left(\sqrt[7]{x}\right)^2}$ 20. $(xy)^{-3/5} = \dfrac{1}{\sqrt[5]{(xy)^3}}$ or $\dfrac{1}{\left(\sqrt[5]{xy}\right)^3}$

22. $4x^{-3/2} = \dfrac{4}{\sqrt{x^3}}$ or $\dfrac{4}{(\sqrt{x})^3}$ 24. $-3x^{2/5}y^{3/5} = -3\sqrt[5]{x^2}\sqrt[5]{y^3} = \sqrt[5]{x^2 y^3}$

26. $9(x - 2y)^{2/3} = 9\sqrt[3]{(x - 2y)^2}$ or $9\left(\sqrt[3]{x - 2y}\right)^2$

28. $(x^3 - 8)^{-5/3} = \dfrac{1}{\sqrt[3]{(x^3 - 8)^5}}$ or $\dfrac{1}{\left(\sqrt[3]{x^3 - 8}\right)^5}$

30. $\sqrt{y^3} = y^{3/2}$

32. $\sqrt[3]{ab^2} = a^{1/3}b^{2/3}$

34. $6\sqrt[5]{(ab)^3} = 6(ab)^{3/5}$ or $6a^{3/5}b^{3/5}$

36. $\dfrac{-2x}{\sqrt[3]{y^2}} = -2xy^{-2/3}$

38. $-\sqrt[3]{(2a - b^2)^2} = -(2a - b^2)^{2/3}$

40. $\dfrac{b^2}{\sqrt[4]{3(a^3 + b)^3}} = 3^{-1/4}b^2(a^3 + b)^{-3/4}$

42. $\sqrt[4]{16^5} = 2^5 = 32$

44. $\sqrt[3]{\left(\dfrac{-125}{64}\right)^2} = \left(\dfrac{-5}{4}\right)^2 = \dfrac{25}{16}$

46. $\sqrt[5]{243x^{10}} = 243^{1/5}x^{10/5} = 3x^2$

48. $-\sqrt{a^{10}b^{36}} = -a^{10/2}b^{36/2} = -a^5b^{18}$

50. $-\sqrt[3]{\dfrac{64}{27}x^6y^{18}} = -\dfrac{4}{3}x^{6/3}y^{18/3} = -\dfrac{4}{3}x^2y^6$

52. $\sqrt[5]{-32x^{25}y^5} = -2x^{25/5}y^{5/5} = -2x^5y$

54. $20^{5/4} \approx 42.295$

56. $\sqrt[5]{-8^3} = \sqrt[5]{-512} \approx -3.482$

58. $123^{-3/2} = \dfrac{1}{\left(\sqrt{123}\right)^3} \approx 0.001$

60. $16.1^{0.29} \approx 2.239$

62. $4x^{1/3}x^{5/6} = 4x^{2/6 + 5/6} = 4x^{7/6}$

64. $\dfrac{xy^2}{5x^{1/4}} = \dfrac{1}{5}x^{4/4 - 1/4}y^2 = \dfrac{1}{5}x^{3/4}y^2$

66. $\left(\dfrac{1}{2}x^{-3/4}\right)\left(4x^{1/4}\right) = 2x^{-2/4} = \dfrac{2}{x^{1/2}}$

68. $(-32y^{-2/3})^{3/5} = (-2)^3y^{(-2/3)(3/5)} = \dfrac{-8}{y^{2/5}}$

70. $\left(\dfrac{a^{9/2}}{8b^{-6}}\right)^{-4/3} = \dfrac{8^{4/3}a^{(9/2)(-4/3)}}{b^{-6(-4/3)}} = \dfrac{16}{a^6b^8}$

72. $\left(\dfrac{a^{2.1}b^{1.1}}{a^3}\right)^{-0.6} = \dfrac{b^{1.1(-0.6)}}{a^{0.9(-0.6)}} = \dfrac{a^{0.54}}{b^{0.66}}$

74. $\dfrac{\left(2x^{-1/2}y\right)^{-3}}{8\left(x^{-3/4}y^6\right)^{4/3}} = \dfrac{2^{-3}x^{3/2}y^{-3}}{8x^{(-3/4)(4/3)}y^{6(4/3)}} = \dfrac{x^{3/2}x^1}{8(8)y^3y^8} = \dfrac{x^{5/2}}{64y^{11}}$

76. $\left(\dfrac{a^3}{b^{4/3}}\right)^{1/4}\left(\dfrac{b^{-2}}{a^2}\right)^{3/4} = \left(\dfrac{a^{3/4}}{b^{1/3}}\right)\left(\dfrac{1}{a^{3/2}b^{3/2}}\right) = \dfrac{1}{a^{3/4}b^{11/6}}$

78. $x^{1/3}(2x^{2/3} - x^{1/3}) = 2x^{1/3}x^{2/3} - x^{1/3}x^{1/3} = 2x - x^{2/3}$

80. $\dfrac{1}{2}y^{-1/3}(y^{2/3} + 3y^{-5/6}) = \dfrac{1}{2}y^{-1/3}y^{2/3} + \dfrac{3}{2}y^{-2/6}y^{-5/6} = \dfrac{1}{2}y^{1/3} + \dfrac{3}{2}y^{-7/6}$

82. $(2x^{1/4} + 1)(x^{1/4} - 1) = 2x^{2/4} + x^{1/4} - 2x^{1/4} - 1 = 2x^{1/2} - x^{1/4} - 1$

84. $(a^{2/3} + 3)^2 = a^{4/3} + 6a^{2/3} + 9$

86. $\dfrac{3b^{-1/4} + \frac{1}{2}b^{5/4}}{2b^{3/4}} = \dfrac{3b^{-1/4}}{2b^{3/4}} + \dfrac{\frac{1}{2}b^{5/4}}{2b^{3/4}} = \dfrac{3}{2}b^{-1} + \dfrac{1}{4}b^{1/2}$

88. $x^{-3.1}(2x^{1.2} + x^{0.4}) = 2x^{1.2-3.1} + x^{0.4-3.1} = 2x^{-1.9} + x^{-2.7}$

90. $(2y^{0.2} + 1)(3y^{-1.8} - 2) = 6y^{-1.6} - 4y^{0.2} + 3y^{-1.8} - 2$

92. population $= 2000(30)^{5/4} \approx 2000(70.21) = 140{,}420$ people

94. distance $= (1.243 \times 10^{-24})^{-1/3}(0.615)^{2/3} \approx 67{,}260{,}440$ miles

96. intensity $= 10^{8.3} = 199{,}526{,}231 \approx 200$ million times as intense

98. air pressure $= (2.79 \times 10^4)(50^{-1.4}) \approx 116.7$ pounds per square inch

100. value $= (2500)2^{-2/3} \approx 2500(0.62996) = \$1{,}574.90$

102. cost $= 1500(1 + 0.065)^{1.5} = 1500(1.065)^{1.5} \approx \$1{,}648.60$

104. $y - y^{2/3} = y^{2/3}(y^{1/3} - 1)$ 106. $y^{3/4} - y^{-1/4} = y^{-1/4}(y - 1)$

108. $(y + 2)^{1/5} - (y + 2)^{-4/5} = (y + 2)^{-4/5}((y + 2) - 1) = (y + 2)^{-4/5}(y + 1)$

110. $x^{1/2}(x - 3)^{-1/2} + x^{-1/2}(x - 3)^{-3/2} = x^{-1/2}(x - 3)^{-3/2}(x(x - 3) + 1) =$

$x^{-1/2}(x - 3)^{-3/2}(x^2 - 3x + 1)$

112. $-z^{-4/3}(z + 2)^{4/3} + \dfrac{4}{3}z^{-1/3}(z + 2)^{1/3} = z^{-4/3}(z + 2)^{1/3}(-(z + 2) + \dfrac{4}{3}z) =$

$z^{-4/3}(z + 2)^{1/3}(\dfrac{1}{3}z - 2)$

Exercise 3.3

2. $\sqrt{50} = \sqrt{5^2}\sqrt{2} = 5\sqrt{2}$ 4. $\sqrt[3]{54} = \sqrt[3]{3^3}\sqrt[3]{2} = 3\sqrt[3]{2}$

6. $-\sqrt[4]{162} = -\sqrt[4]{3^4}\sqrt[4]{2} = -3\sqrt[4]{2}$ 8. $\sqrt{800{,}000} = \sqrt{400^2}\sqrt{5} = 400\sqrt{5}$

10. $\sqrt[3]{24{,}000} = \sqrt[3]{20^3}\sqrt[3]{3} = 20\sqrt[3]{3}$ 12. $\sqrt[4]{\dfrac{80}{625}} = \sqrt[4]{\dfrac{2^4}{5^4}}\sqrt[4]{5} = \dfrac{2}{5}\sqrt[4]{5}$

14. $\sqrt[3]{y^{16}} = \sqrt[3]{y^{15}}\sqrt[3]{y} = y^5\sqrt[3]{y}$ 16. $\sqrt{12t^5} = \sqrt{2^2t^4}\sqrt{3t} = 2t^2\sqrt{3t}$

18. $\sqrt[3]{81a^{12}b^8} = \sqrt[3]{3^3a^{12}b^6}\sqrt[3]{3b^2} = 3a^4b^2\sqrt[3]{3b^2}$

20. $-\sqrt[6]{256k^7u^{12}v^{15}} = -\sqrt[6]{2^6k^6u^{12}v^{12}}\sqrt[6]{4kv^3} = -2ku^2v^2\sqrt[6]{4kv^3}$

22. $\sqrt{3w^3}\sqrt{27w^3} = \sqrt{81w^6} = 9w^3$ 24. $-\sqrt[4]{2m^3}\sqrt[4]{8m} = -\sqrt[4]{16m^4} = -2m$

26. $\sqrt{9y^2 + 18} = \sqrt{9(y^2 + 2)} = 3\sqrt{y^2 + 2}$

28. $\sqrt[3]{b^9 - 27b^{27}} = \sqrt[3]{b^9(1 - 27b^{18})} = b^3\sqrt[3]{1 - 27b^{18}}$

30. $\sqrt{\dfrac{-16c^{15}b^7}{-169n^6}} = \sqrt{\dfrac{16c^{14}b^6}{169n^6}}\sqrt{bc} = \dfrac{4c^7b^3}{13n^3}\sqrt{bc}$

32. $\sqrt[6]{\dfrac{192k^{13}}{m^{18}}} = \sqrt[6]{\dfrac{64k^{12}}{m^{18}}}\sqrt[6]{3k} = \dfrac{4k^2}{m^3}\sqrt[6]{3k}$ 34. $\dfrac{10}{\sqrt{5}} = \dfrac{10\sqrt{5}}{\sqrt{5}\sqrt{5}} = \dfrac{10\sqrt{5}}{5} = 2\sqrt{5}$

36. $\dfrac{-\sqrt{5}}{\sqrt{6}} = \dfrac{-\sqrt{5}\sqrt{6}}{\sqrt{6}\sqrt{6}} = \dfrac{-\sqrt{30}}{6}$ 38. $\sqrt{\dfrac{27x}{20}} = \dfrac{3\sqrt{3x}}{2\sqrt{5}} = \dfrac{3\sqrt{3x}\sqrt{5}}{2\sqrt{5}\sqrt{5}} = \dfrac{3\sqrt{15x}}{10}$

40. $\sqrt{\dfrac{5p}{q}} = \dfrac{\sqrt{5p}\sqrt{q}}{\sqrt{q}\sqrt{q}} = \dfrac{\sqrt{5pq}}{q}$ 42. $\dfrac{6\sqrt{2}}{\sqrt{3v}} = \dfrac{6\sqrt{2}\sqrt{3v}}{\sqrt{3v}\sqrt{3v}} = \dfrac{6\sqrt{6v}}{3v} = \dfrac{2\sqrt{6v}}{v}$

44. $\dfrac{-8y\sqrt{21y^5}}{3\sqrt{10t}} = \dfrac{-8y^3\sqrt{21y}\sqrt{10t}}{3\sqrt{10t}\sqrt{10t}} = \dfrac{-8y^3\sqrt{210ty}}{30t} = \dfrac{-4y^3\sqrt{210ty}}{15t}$

46. $\dfrac{1}{\sqrt[4]{y^3}} = \dfrac{1\sqrt[4]{y}}{\sqrt[4]{y^3}\sqrt[4]{y}} = \dfrac{\sqrt[4]{y}}{\sqrt[4]{y^4}} = \dfrac{\sqrt[4]{y}}{y}$ 48. $\sqrt[4]{\dfrac{2}{3x}} = \dfrac{\sqrt[4]{2}\sqrt[4]{27x^3}}{\sqrt[4]{3x}\sqrt[4]{27x^3}} = \dfrac{\sqrt[4]{54x^3}}{\sqrt[4]{81x^4}} = \dfrac{\sqrt[4]{54x^3}}{3x}$

50. $\sqrt[4]{\dfrac{x}{8y^3}} = \dfrac{\sqrt[4]{x}\sqrt[4]{2y}}{\sqrt[4]{8y^3}\sqrt[4]{2y}} = \dfrac{\sqrt[4]{2xy}}{\sqrt[4]{16y^4}} = \dfrac{\sqrt[4]{2xy}}{2y}$

52. $\sqrt[5]{\dfrac{2}{9y^2}} = \dfrac{\sqrt[5]{2}\sqrt[5]{27y^3}}{\sqrt[5]{9y^2}\sqrt[5]{27y^3}} = \dfrac{\sqrt[5]{54y^3}}{\sqrt[5]{243y^5}} = \dfrac{\sqrt[5]{54y^3}}{3y}$

31

54. $\dfrac{15x^4}{\sqrt[3]{5x}} = \dfrac{15x^4\sqrt[3]{25x^2}}{\sqrt[3]{5x}\,\sqrt[3]{25x^2}} = \dfrac{15x^4\sqrt[3]{25x^2}}{\sqrt[3]{125x^3}} = \dfrac{15x^4\sqrt[3]{25x^2}}{5x} = 3x^3\sqrt[3]{25x^2}$

56. $\dfrac{x+y}{\sqrt[4]{(x+y)^2}} = \dfrac{(x+y)\sqrt[4]{(x+y)^2}}{\sqrt[4]{(x+y)^2}\,\sqrt[4]{(x+y)^2}} = \dfrac{(x+y)\sqrt[4]{(x+y)^2}}{(x+y)} = \sqrt[4]{(x+y)^2} \ \text{or} \ \sqrt{x+y}$

58. $\dfrac{\sqrt{x}\sqrt{xy^3}}{\sqrt{y}} = \sqrt{\dfrac{x^2y^3}{y}} = \sqrt{x^2y^2} = xy$

60. $\dfrac{\sqrt{45x^3}\sqrt{y^3}}{\sqrt{5y}} = \sqrt{\dfrac{45x^3y^3}{5y}} = \sqrt{9x^3y^2} = 3xy\sqrt{x}$

62. $\dfrac{\sqrt[3]{16r^4}}{\sqrt[3]{4t^3}} = \sqrt[3]{\dfrac{16r^4}{4t^3}} = \sqrt[3]{\dfrac{4r^4}{t^3}} = \dfrac{r\sqrt[3]{4r}}{t}$
64. $\dfrac{\sqrt[5]{x^2}\sqrt[5]{y^3}}{\sqrt[5]{xy^2}} = \sqrt[5]{\dfrac{x^2y^3}{xy^2}} = \sqrt[5]{xy}$

66. $\sqrt[6]{2^2} = 2^{2/6} = 2^{1/3} = \sqrt[3]{2}$
68. $\sqrt[8]{5^2} = 5^{2/8} = 5^{1/4} = \sqrt[4]{5}$

70. $\sqrt[10]{32} = 2^{5/10} = 2^{1/2} = \sqrt{2}$
72. $\sqrt[9]{y^3} = y^{3/9} = y^{1/3} = \sqrt[3]{y}$

74. $\dfrac{\sqrt[4]{8ab^3}\sqrt[4]{8a^2b^3}}{\sqrt[4]{2a^5}} = \sqrt[4]{\dfrac{32b^6}{a^2}} = \dfrac{2b\sqrt[4]{2b^2}\sqrt[4]{a^2}}{\sqrt[4]{a^2}\sqrt[4]{a^2}} = \dfrac{2b\sqrt[4]{2a^2b^2}}{a}$

76. $\dfrac{6xy^3}{\sqrt[5]{4x^3y^4}\sqrt[5]{2x^6y^3}} = \dfrac{6xy^3}{\sqrt[5]{8x^9y^7}} = \dfrac{6xy^3\sqrt[5]{4xy^3}}{xy\sqrt[5]{8x^4y^2}\sqrt[5]{4xy^3}} = \dfrac{6xy^3\sqrt[5]{4xy^3}}{xy(2xy)} = \dfrac{3y\sqrt[5]{4xy^3}}{x}$

78. $\sqrt[8]{81x^6z^5}\sqrt[8]{x^4z^{15}} = \sqrt[8]{81x^{10}z^{20}} = xz^2\sqrt[8]{3^4x^2z^4} = xz^2(3^{4/8}x^{2/8}z^{4/8}) =$

$xz^2(3^{2/4}x^{1/4}z^{2/4}) = xz^2\sqrt[4]{9xz^2}$

80. $\sqrt{15y(2y-3)}\,\sqrt{30y(2y-3)^3} = \sqrt{450y^2(2y-3)^4} = 15y(2y-3)^2\sqrt{2}$

82. $\sqrt[3]{9ab(a^2+3)^2}\sqrt[3]{12a^2b^4(a^2+3)^2} = \sqrt[3]{108a^3b^5(a^2+3)^4} = 3ab(a^2+3)\sqrt[3]{4b^2(a^2+3)}$

84. $\sqrt{12(x+2)^3} = \sqrt{4(x+2)^2}\,\sqrt{3(x+2)} = 2(x+2)\sqrt{3(x+2)}$

86. $\sqrt{y^5(x-1)^3} = \sqrt{y^4(x-1)^2}\,\sqrt{y(x-1)} = y^2(x-1)\sqrt{y(x-1)}$

88. $\sqrt{\dfrac{(x+2)^5}{x^3y}} = \dfrac{\sqrt{(x+2)^4}\,\sqrt{x+2}\,\sqrt{xy}}{x\sqrt{xy}\sqrt{xy}} = \dfrac{(x+2)^2\sqrt{(x+2)xy}}{x^2y}$

90. $\sqrt[3]{\dfrac{(y+1)^4}{x^2y}} = \dfrac{(y+1)\sqrt[3]{y+1}\,\sqrt[3]{xy^2}}{\sqrt[3]{x^2y}\,\sqrt[3]{xy^2}} = \dfrac{(y+1)\sqrt[3]{xy^2(y+1)}}{xy}$

92. $\sqrt[3]{(2y^4-y^3)(y+2)^5} = \sqrt[3]{y^3(y+2)^3}\,\sqrt[3]{(2y-1)(y+2)^2} = y(y+2)\sqrt[3]{(2y-1)(y+2)}$

94. $\dfrac{\sqrt{4(y+2)^2}}{\sqrt{4y}\,\sqrt{y^3+2y^2}} = \sqrt{\dfrac{4(y+2)^2}{4y(y^2)(y+2)}} = \dfrac{\sqrt{y+2}\sqrt{y}}{y\sqrt{y}\sqrt{y}} = \dfrac{\sqrt{y(y+2)}}{y^2}$

96. $\sqrt{9x^2y^4} = 9\,|x|\,y^2$ 98. $\sqrt{4x^2-4x+1} = \sqrt{(2x-1)^2} = |2x-1|$

100. $\dfrac{3}{\sqrt{x^4+2x^2y^2+y^4}} = \dfrac{3}{\sqrt{(x^2+y^2)^2}} = \dfrac{3}{x^2+y^2}$ (since $x^2+y^2 \geq 0$)

Exercise 3.4

2. $5\sqrt{2}-3\sqrt{2} = (5-3)\sqrt{2} = 2\sqrt{2}$ 4. $\sqrt{75}+2\sqrt{3} = 5\sqrt{3}+2\sqrt{3} = (5+2)\sqrt{3} = 7\sqrt{3}$

6. $\sqrt{8y}-\sqrt{18y} = 2\sqrt{2y}-3\sqrt{2y} = -\sqrt{2y}$

8. $\sqrt[3]{81}+2\sqrt[3]{24}-3\sqrt[3]{3} = 3\sqrt[3]{3}+2(2)\sqrt[3]{3}-3\sqrt[3]{3} = 4\sqrt[3]{3}$

10. $6\sqrt[3]{32}-3\sqrt{32}+\sqrt[3]{128}-2\sqrt{128} = 12\sqrt[3]{4}-12\sqrt{2}+4\sqrt[3]{2}-16\sqrt{2} = 12\sqrt[3]{4}+4\sqrt[3]{2}-28\sqrt{2}$

12. $2\sqrt{8y^2z}-3\sqrt{9yz^2}+3\sqrt{32y^2z} = 4y\sqrt{2z}-9z\sqrt{y}+12y\sqrt{2z} = 16y\sqrt{2z}-9z\sqrt{y}$

14. $x^2\sqrt[4]{48xy}-\dfrac{3x}{y}\sqrt[4]{3x^5y^5}-7\sqrt[4]{3x^9y} = 2x^2\sqrt[4]{3xy}-3x^2\sqrt[4]{3xy}-7x^2\sqrt[4]{3xy} = -8x^2\sqrt[4]{3xy}$

16. $\dfrac{\sqrt{5}}{3}-\sqrt{80} = \dfrac{1}{3}\sqrt{5}-4\sqrt{5} = -\dfrac{11}{3}\sqrt{5}$

18. $\dfrac{2}{3}\sqrt[3]{3y}+\sqrt[3]{24y} = \dfrac{2}{3}\sqrt[3]{3y}+3\sqrt[3]{3y} = \dfrac{11}{3}\sqrt[3]{3y}$

20. $\dfrac{ab\sqrt{2}}{3}+\dfrac{3ab}{\sqrt{2}} = \dfrac{ab\sqrt{2}}{3}+\dfrac{3ab\sqrt{2}}{2} = \dfrac{2ab\sqrt{2}}{6}+\dfrac{9ab\sqrt{2}}{6} = \dfrac{11ab\sqrt{2}}{6}$

22. $5(2-\sqrt{7}) = 10-5\sqrt{7}$ 24. $\sqrt{3}(\sqrt{12}-\sqrt{15}) = \sqrt{36}-\sqrt{45} = 6-\sqrt{9}\sqrt{5} = 6-3\sqrt{5}$

26. $\sqrt[3]{3}(2\sqrt[3]{18} + \sqrt[3]{36}) = 2\sqrt[3]{54} + \sqrt[3]{108} = 2\sqrt[3]{27}\sqrt[3]{2} + \sqrt[3]{27}\sqrt[3]{4} = 6\sqrt[3]{2} + 3\sqrt[3]{4}$

28. $\sqrt{3y}(\sqrt{6y} - \sqrt{18}) = \sqrt{18y^2} - \sqrt{54y} = \sqrt{9y^2}\sqrt{2} - \sqrt{9}\sqrt{6y} = 3y\sqrt{2} - 3\sqrt{6y}$

30. $\sqrt{4-x}(\sqrt{4x} + \sqrt{4-x}) = \sqrt{4x(4-x)} + \sqrt{4-x}\sqrt{4-x} = \sqrt{16x - 4x^2} + 4 - x$

32. $\sqrt[3]{(x+2)^2}(\sqrt[3]{(x+2)^2} + \sqrt[3]{2x}) = \sqrt[3]{(x+2)^4} + \sqrt[3]{2x(x+2)^2} =$

$\sqrt[3]{(x+2)^3}\sqrt[3]{x+2} + \sqrt[3]{2x(x+2)^2} = (x+2)\sqrt[3]{x+2} + \sqrt[3]{2x(x+2)^2}$

34. $(2 + \sqrt{x})(2 - \sqrt{x}) = 2^2 - (\sqrt{x})^2 = 4 - x$

36. $(\sqrt{3} - \sqrt{5})(2\sqrt{3} + \sqrt{5}) = 2(\sqrt{3})^2 + \sqrt{15} - 2\sqrt{15} - (\sqrt{5})^2 = 2(3) - \sqrt{15} - 5 = 1 - \sqrt{15}$

38. $(\sqrt{2} - 2\sqrt{3})^2 = (\sqrt{2})^2 - 2\sqrt{2}(2\sqrt{3}) + (2\sqrt{3})^2 = 2 - 4\sqrt{6} + 12 = 14 - 4\sqrt{6}$

40. $(\sqrt{5x} - 2\sqrt{y})(2\sqrt{5x} - 3\sqrt{y}) = 2(\sqrt{5x})^2 - 3\sqrt{5xy} - 4\sqrt{5xy} + 6(\sqrt{y})^2 =$

$2(5x) - 7\sqrt{5xy} + 6y = 10x + 6y - 7\sqrt{5xy}$

42. $(\sqrt{2a} - 2\sqrt{b})(\sqrt{2a} + 2\sqrt{b}) = (\sqrt{2a})^2 - (2\sqrt{b})^2 = 2a - 4b$

44. $\sqrt[3]{6\sqrt{6}} = 6^{1/3}6^{1/2} = 6^{5/6} = \sqrt[6]{6^5} = \sqrt[6]{7776}$

46. $\sqrt[3]{6\sqrt{3}} = 6^{1/3}3^{1/2} = 6^{2/6}3^{3/6} = 36^{1/6}27^{1/6} = (36 \cdot 27)^{1/6} = \sqrt[6]{972}$

48. $\sqrt[3]{2x^2}\sqrt[5]{x^2} = 2^{1/3}(x^2)^{1/3}(x^2)^{1/5} = 2^{1/3}(x^2)^{1/3+1/5} = 2^{5/15}x^{16/15} = 32^{1/15}x^1x^{1/15} =$

$x \sqrt[15]{32x}$

50. $\sqrt{xy}\sqrt[5]{x^2y^3} = x^{1/2}y^{1/2}x^{2/5}y^{3/5} = x^{5/10}y^{5/10}x^{4/10}y^{6/10} = x^{9/10}y^{11/10} = \sqrt[10]{x^9y^{11}} =$

$y \sqrt[10]{x^9y}$

52. $5 + 10\sqrt{2} = 5(1 + 2\sqrt{2})$ 54. $5\sqrt{5} - \sqrt{25} = 5(\sqrt{5} - 1)$

56. $3 + \sqrt{18x} = 3 + 3\sqrt{2x} = 3(1 + \sqrt{2x})$

58. $\sqrt{12} - 2\sqrt{6} = 2\sqrt{3} - 2\sqrt{3}\sqrt{2} = 2\sqrt{3}(1 - \sqrt{2})$

60. $a\sqrt{5b} - \sqrt{3ab} = \sqrt{b}(a\sqrt{5} - \sqrt{3a})$

34

62. $6y + \sqrt{18y} = 6\sqrt{y}\sqrt{y} + 3\sqrt{2y} = 3\sqrt{y}(2\sqrt{y} + \sqrt{2})$

64. $\dfrac{6 + 2\sqrt{5}}{2} = \dfrac{2(3 + \sqrt{5})}{2} = 3 + \sqrt{5}$ 66. $\dfrac{8 - 2\sqrt{12}}{4} = \dfrac{8 - 4\sqrt{3}}{4} = \dfrac{4(2 - \sqrt{3})}{4} = 2 - \sqrt{3}$

68. $\dfrac{xy - x\sqrt{xy^2}}{xy} = \dfrac{xy - xy\sqrt{x}}{xy} = \dfrac{xy(1 - \sqrt{x})}{xy} = 1 - \sqrt{x}$

70. $\dfrac{\sqrt{x} - y\sqrt{x^3}}{\sqrt{x}} = \dfrac{\sqrt{x} - xy\sqrt{x}}{\sqrt{x}} = \dfrac{\sqrt{x}(1 - xy)}{\sqrt{x}} = 1 - xy$

72. $\dfrac{12b - \sqrt{12ab}}{6\sqrt{b}} = \dfrac{12b - 2\sqrt{3ab}}{6\sqrt{b}} = \dfrac{2\sqrt{b}(6\sqrt{b} - \sqrt{3a})}{6\sqrt{b}} = \dfrac{6\sqrt{b} - \sqrt{3a}}{3}$

74. $\dfrac{4x + \sqrt[4]{x^3}}{\sqrt[4]{x}} = \dfrac{4\sqrt[4]{x^4} + \sqrt[4]{x^3}}{\sqrt[4]{x}} = \dfrac{\sqrt[4]{x}(4\sqrt[4]{x^3} + \sqrt[4]{x^2})}{\sqrt[4]{x}} = 4\sqrt[4]{x^3} + \sqrt[4]{x^2} = 4\sqrt[4]{x^3} + \sqrt{x}$

76. $\dfrac{1}{2 - \sqrt{2}} = \dfrac{1(2 + \sqrt{2})}{(2 - \sqrt{2})(2 + \sqrt{2})} = \dfrac{2 + \sqrt{2}}{4 - 2} = \dfrac{2 + \sqrt{2}}{2}$

78. $\dfrac{2}{4 - \sqrt{5}} = \dfrac{2(4 + \sqrt{5})}{(4 - \sqrt{5})(4 + \sqrt{5})} = \dfrac{2(4 + \sqrt{5})}{16 - 5} = \dfrac{8 + 2\sqrt{5}}{11}$

80. $\dfrac{y}{\sqrt{3} - y} = \dfrac{y(\sqrt{3} + y)}{(\sqrt{3} - y)(\sqrt{3} + y)} = \dfrac{y\sqrt{3} + y^2}{3 - y^2}$

82. $\dfrac{\sqrt{x} + \sqrt{y}}{\sqrt{x} - \sqrt{y}} = \dfrac{(\sqrt{x} + \sqrt{y})(\sqrt{x} + \sqrt{y})}{(\sqrt{x} - \sqrt{y})(\sqrt{x} + \sqrt{y})} = \dfrac{x + 2\sqrt{xy} + y}{x - y}$

84. $\dfrac{\sqrt{3}}{2\sqrt{3} - 3\sqrt{2}} = \dfrac{\sqrt{3}(2\sqrt{3} + 3\sqrt{2})}{(2\sqrt{3} - 3\sqrt{2})(2\sqrt{3} + 3\sqrt{2})} = \dfrac{3(2) + 3\sqrt{6}}{12 - 18} = \dfrac{3(2 + \sqrt{6})}{-6} = -\dfrac{2 + \sqrt{6}}{2}$

86. $\dfrac{\sqrt{6b} + 6\sqrt{a}}{3\sqrt{b} - \sqrt{3a}} = \dfrac{(\sqrt{6b} + 6\sqrt{a})(3\sqrt{b} + \sqrt{3a})}{(3\sqrt{b} - \sqrt{3a})(3\sqrt{b} + \sqrt{3a})} = \dfrac{3b\sqrt{6} + \sqrt{18ab} + 18\sqrt{ab} + 6a\sqrt{3}}{9b - 3a} =$

$\dfrac{3(b\sqrt{6} + \sqrt{2ab} + 6\sqrt{ab} + 2a\sqrt{3})}{3(3b - a)} = \dfrac{b\sqrt{6} + \sqrt{2ab} + 6\sqrt{ab} + 2a\sqrt{3}}{3b - a}$

88. $\dfrac{1 - \sqrt{x + 1}}{\sqrt{x + 1}} = \dfrac{(1 - \sqrt{x + 1})(1 + \sqrt{x + 1})}{\sqrt{x + 1}(1 + \sqrt{x + 1})} = \dfrac{1 - (x + 1)}{\sqrt{x + 1} + (x + 1)} = \dfrac{-x}{\sqrt{x + 1} + x + 1}$

90.

$$\frac{\sqrt{x-1}+\sqrt{x}}{\sqrt{x-1}-\sqrt{x}} = \frac{(\sqrt{x-1}+\sqrt{x})(\sqrt{x-1}-\sqrt{x})}{(\sqrt{x-1}-\sqrt{x})(\sqrt{x-1}-\sqrt{x})} = \frac{(x-1)-x}{(x-1)-2\sqrt{x}\sqrt{x-1}+x} =$$

$$\frac{-1}{2x-1-2\sqrt{x^2-x}}$$

92.

$$\frac{\sqrt{x}}{x} - \frac{x}{\sqrt{x}} = \frac{\sqrt{x}}{x} - \frac{x\sqrt{x}}{\sqrt{x}\sqrt{x}} = \frac{\sqrt{x}-x\sqrt{x}}{x}$$

94.

$$\sqrt{x^2-2} - \frac{x^2+1}{\sqrt{x^2-2}} = \frac{\sqrt{x^2-2}\sqrt{x^2-2}}{\sqrt{x^2-2}} - \frac{x^2+1}{\sqrt{x^2-2}} = \frac{x^2-2-(x^2+1)}{\sqrt{x^2-2}} = \frac{-3\sqrt{x^2-2}}{\sqrt{x^2-2}\sqrt{x^2-2}} =$$

$$\frac{-3\sqrt{x^2-2}}{x^2-2}$$

96.

$$\frac{x}{\sqrt{x^2-1}} + \frac{\sqrt{x^2-1}}{x} = \frac{x^2}{x\sqrt{x^2-1}} + \frac{\sqrt{x^2-1}\sqrt{x^2-1}}{x\sqrt{x^2-1}} = \frac{x^2+x^2-1}{x\sqrt{x^2-1}} = \frac{(2x^2-1)\sqrt{x^2-1}}{x\sqrt{x^2-1}\sqrt{x^2-1}} =$$

$$\frac{(2x^2-1)\sqrt{x^2-1}}{x^3-x}$$

98.

$$\frac{\sqrt{x}+\dfrac{1}{\sqrt{x}}}{x+1} = \frac{\sqrt{x}\left(\sqrt{x}+\dfrac{1}{\sqrt{x}}\right)}{\sqrt{x}\,(x+1)} = \frac{x+1}{\sqrt{x}\,(x+1)} = \frac{1}{\sqrt{x}} = \frac{\sqrt{x}}{x}$$

100.

$$\frac{\dfrac{1}{\sqrt{x+2}}+\sqrt{x}}{\dfrac{\sqrt{x}}{\sqrt{x+2}}} = \frac{\sqrt{x+2}\left(\dfrac{1}{\sqrt{x+2}}+\sqrt{x}\right)}{\dfrac{\sqrt{x+2}\sqrt{x}}{\sqrt{x+2}}} = \frac{1+\sqrt{x(x+2)}}{\sqrt{x}} = \frac{\sqrt{x}\,(1+\sqrt{x(x+2)})}{\sqrt{x}\sqrt{x}} =$$

$$\frac{\sqrt{x}+x\sqrt{x+2}}{x}$$

102.

$$\frac{\dfrac{x}{\sqrt{x^2-1}}-\sqrt{x^2-1}}{x^2\sqrt{x^2-1}} = \frac{\sqrt{x^2-1}\left(\dfrac{x}{\sqrt{x^2-1}}-\sqrt{x^2-1}\right)}{x^2\sqrt{x^2-1}\sqrt{x^2-1}} = \frac{x-(x^2-1)}{x^2(x^2-1)} = \frac{-x^2+x+1}{x^4-x^2}$$

Exercise 3.5

2. $\sqrt{-9} = \sqrt{-1}\sqrt{9} = i\sqrt{9} = 3i$

4. $\sqrt{-50} = \sqrt{-1}\sqrt{25}\sqrt{2} = 5i\sqrt{2}$

6. $4\sqrt{-18} = 4\sqrt{-1}\sqrt{9}\sqrt{2} = 12i\sqrt{2}$

8. $2\sqrt{-40} = 2\sqrt{-1}\sqrt{4}\sqrt{10} = 4i\sqrt{10}$

10. $7\sqrt{-81} = 7\sqrt{-1}\sqrt{81} = 63i$

12. $-3\sqrt{-75} = -3\sqrt{-1}\sqrt{25}\sqrt{3} = -15i\sqrt{3}$

14. $5 - 3\sqrt{-1} = 5 - 3i$

16. $5\sqrt{-12} - 1 = 5\sqrt{-1}\sqrt{4}\sqrt{3} - 1 = -1 + 10i\sqrt{3}$

18. $\sqrt{20} - \sqrt{-20} = \sqrt{4}\sqrt{5} - \sqrt{-1}\sqrt{4}\sqrt{5} = 2\sqrt{5} - 2i\sqrt{5}$

20. $(2 - i) + (3 - 2i) = (2 + 3) + (-1 + -2)i = 5 - 3i$

22. $(2 + i) - (4 - 2i) = (2 - 4) + (1 - (-2))i = -2 + 3i$

24. $(2 - 6i) - 3 = (2 - 3) + (-6 - 0)i = -1 - 6i$

26. $(1 - 3i)(4 - 5i) = 4 - 5i - 12i + 15i^2 = 4 - 17i - 15 = -11 - 17i$

28. $(-3 - i)(2 - 3i) = -6 + 9i - 2i + 3i^2 = -6 + 7i - 3 = -9 + 7i$

30. $(7 + 3i)(-2 - 3i) = -14 - 21i - 6i - 9i^2 = -14 - 27i + 9 = -5 - 27i$

32. $(2 + 3i)^2 = 4 + 2(6i) + 9i^2 = 4 + 12i - 9 = -5 + 12i$

34. $(1 - 2i)(1 + 2i) = 1 + 2i - 2i - 4i^2 = 1 + 4 = 5$

36. $\dfrac{-2}{5i} = \dfrac{-2i}{5i^2} = \dfrac{-2i}{-5} = \dfrac{2}{5}i$

38. $\dfrac{4 + 2i}{3i} = \dfrac{(4 + 2i)i}{3i^2} = \dfrac{4i + 2i^2}{-3} = \dfrac{4i - 2}{-3} = \dfrac{2}{3} - \dfrac{4}{3}i$

40. $\dfrac{-3}{2 + i} = \dfrac{-3(2 - i)}{(2 + i)(2 - i)} = \dfrac{-6 + 3i}{4 - i^2} = \dfrac{-6 + 3i}{4 + 1} = \dfrac{-6}{5} + \dfrac{3}{5}i$

42. $\dfrac{3 - i}{1 + i} = \dfrac{(3 - i)(1 - i)}{(1 + i)(1 - i)} = \dfrac{3 - 3i - i + i^2}{1 - i^2} = \dfrac{3 - 4i - 1}{1 + 1} = \dfrac{2 - 4i}{2} = 1 - 2i$

44. $\dfrac{6 + i}{2 - 5i} = \dfrac{(6 + i)(2 + 5i)}{(2 - 5i)(2 + 5i)} = \dfrac{12 + 30i + 2i + 5i^2}{4 - 25i^2} = \dfrac{12 + 32i - 5}{4 + 25} = \dfrac{7 + 32i}{29} = \dfrac{7}{29} + \dfrac{32}{29}i$

46. $\dfrac{-4 - 3i}{2 + 7i} = \dfrac{(-4 - 3i)(2 - 7i)}{(2 + 7i)(2 - 7i)} = \dfrac{-8 + 28i - 6i + 21i^2}{4 - 49i^2} = \dfrac{-8 + 22i - 21}{4 + 49} = \dfrac{-29}{53} + \dfrac{22}{53}i$

48. $\sqrt{-9}(3 + \sqrt{-16}) = \sqrt{-1}\sqrt{9}(3 + \sqrt{-1}\sqrt{16}) = 3i(3 + 4i) = 9i + 12i^2 = -12 + 9i$

50. $(4 - \sqrt{-2})(3 + \sqrt{-2}) = (4 - \sqrt{-1}\sqrt{2})(3 + \sqrt{-1}\sqrt{2}) = (4 - i\sqrt{2})(3 + i\sqrt{2}) = $

$12 + 4i\sqrt{2} - 3i\sqrt{2} - 2i^2 = 12 + i\sqrt{2} + 2 = 14 + i\sqrt{2}$

52. $\dfrac{-1}{\sqrt{-25}} = \dfrac{-1}{\sqrt{-1}\sqrt{25}} = \dfrac{-1}{5i} = \dfrac{-1i}{5i^2} = \dfrac{-i}{-5} = \dfrac{1}{5}i$

54. $\dfrac{1 + \sqrt{-2}}{3 - \sqrt{-2}} = \dfrac{1 + \sqrt{-1}\sqrt{2}}{3 - \sqrt{-1}\sqrt{2}} = \dfrac{1 + i\sqrt{2}}{3 - i\sqrt{2}} = \dfrac{(1 + i\sqrt{2})(3 + i\sqrt{2})}{(3 - i\sqrt{2})(3 + i\sqrt{2})} = $

$\dfrac{3 + i\sqrt{2} + 3i\sqrt{2} + 2i^2}{9 - 2i^2} = \dfrac{3 + 4i\sqrt{2} - 2}{9 + 2} = \dfrac{1 + 4i\sqrt{2}}{11} = \dfrac{1}{11} + \dfrac{4}{11}i\sqrt{2}$

37

56. $\sqrt{x + 3}$ is real if $x + 3 \geq 0$ or $x \geq -3$; it is imaginary if $x + 3 < 0$ or $x < -3$.

58. a. $i^{-1} = \frac{1}{i} = \frac{i}{i^2} = \frac{i}{-1} = -i$ b. $i^{-2} = \frac{1}{i^2} = \frac{1}{-1} = -1$

c. $i^{-3} = \frac{1}{i^3} = \frac{i}{i^4} = \frac{i}{1} = i$ d. $i^{-6} = \frac{1}{i^6} = \frac{1}{i^2 i^4} = \frac{1}{-1} = -1$

60. $2(2 - i)^2 - (2 - i) + 2 = 2(4 - 4i + i^2) - (2 - i) + 2 = 2(4 - 4i - 1) - (2 - i) + 2 =$

$6 - 8i - 2 + i + 2 = 6 - 7i$

CHAPTER 4

Exercise 4.1

2. $2 + 5x = 37$

$5x = 35$

$x = 7$

the solution is 7

4. $2(z - 3) = 15$

$2z - 6 = 15$

$2z = 21$

$z = \dfrac{21}{2}$

the solution is $\dfrac{21}{2}$

6. $5y - (y + 1) = 14$

$5y - y - 1 = 14$

$4y = 15$

$y = \dfrac{15}{4}$

the solution is $\dfrac{15}{4}$

8. $3[3x - 2(x - 3) + 1] = 8$

$3[3x - 2x + 6 + 1] = 8$

$3x + 21 = 8$

$3x = -13$

$x = \dfrac{-13}{3}$

the solution is $\dfrac{-13}{3}$

10. $-3[2y - (y - 2)] = 2(y + 3)$

$-3[2y - y + 2] = 2y + 6$

$-3y - 6 = 2y + 6$

$-5y = 12$

$y = \dfrac{-12}{5}$

the solution is $\dfrac{-12}{5}$

12. $(z + 2)^2 - z^2 = 8$

$z^2 + 4z + 4 - z^2 = 8$

$4z = 4$

$z = 1$

the solution is 1

14. $2 + (x - 2)(x + 3) = x^2 + 12$

$2 + x^2 + x - 6 = x^2 + 12$

$x - 4 = 12$

$x = 16$

the solution is 16

16. $2[2x + 3(x + 1)^2 - x^2] = 4x^2 + 1$

$2[2x + 3x^2 + 6x + 3 - x^2] = 4x^2 + 1$

$16x + 4x^2 + 6 = 4x^2 + 1$

$16x = -5$

$x = \dfrac{-5}{16}$

the solution is $\dfrac{-5}{16}$

18. $2(x + 1)^2 + 2x(x + 1) = 2 + (2x - 1)^2$

$2x^2 + 4x + 2 + 2x^2 + 2x = 2 + 4x^2 - 4x + 1$

$4x^2 + 6x + 2 = 4x^2 - 4x + 3$

$10x = 5$

$x = \dfrac{1}{2}$

the solution is $\dfrac{1}{2}$

20. $0.60(y + 2) = 3.60$

       $10[0.6(y + 2)] = 10(3.6)$

       $6y + 12 = 36$

       $6y = 24$

       $y = 4$

       the solution is 4

22. $0.12y + 0.08(y + 10,000) = 12,000$

       $100[0.12y + 0.08y + 800] = 1,200,000$

       $12y + 8y + 80,000 = 1,200,000$

       $20y = 1,120,000$

       $y = 56,000$

       the solution is 56,000

24. $0.10x + 0.12(x + 4,000) = 920$

       $100[0.10x + 0.12x + 480] = 92,000$

       $10x + 12x + 48,000 = 92,000$

       $22x = 44,000$

       $x = 2,000$

       the solution is 2,000

26. $4.8 - 1.3x = 0.7x + 2.1$

       $10[4.8 - 1.3x] = 10[0.7x + 2.1]$

       $48 - 13x = 7x + 21$

       $-20x = -27$

       $x = \dfrac{27}{20}$

       the solution is $\dfrac{27}{20}$

28. $6.4872x - 0.2183 = 2.1847x + 3.4629$

       $10,000[6.4872x - 0.2183] = 10,000[2.1847x + 3.4629]$

       $64,872x - 2,183 = 21,847x + 34,629$

       $43,025x = 36,812$

       $x = \dfrac{36,812}{43,025} = 0.8556$

       the solution is $\dfrac{36,812}{43,025}$ or 0.8556

30. $1.187x + 4.296 = 11.091 + 6.215(x - 7.825)$

       $8,000[1.187x + 4.296] = 8,000[11.091 + 6.215(x - 7.825)]$

       $9,496x + 34,368 = 88,728 + 49,720(x - 7.825)$

       $9,496x + 34,368 = 88,728 + 49,720x - 389,059$

       $-40,224x = -334,699$

       $x = \dfrac{334,699}{40,224}$

       the solution is $\dfrac{334,699}{40,224}$ or 8.32088

32. $V = lwh$

$$h = \frac{V}{lw}$$

34. $S = 3\pi d + \pi a$

$S - \pi a = 3\pi d$

$$d = \frac{S - \pi a}{3\pi}$$

36. $V = k + gt$

$V - k = gt$

$$t = \frac{V - k}{g}$$

38. $A = P(1 + rt)$

$A = P + Prt$

$A - P = Prt$

$$t = \frac{A - P}{Pr}$$

40. $S = a + (n - 1)d$

$S = a + nd - d$

$S - a + d = nd$

$$n = \frac{S - a + d}{d}$$

42. $A = 2w^2 + 4lw$

$A - 2w^2 = 4lw$

$$l = \frac{A - 2w^2}{4w}$$

44. $S = 2(ab + bc + ac)$

$S = 2ab + 2bc + 2ac$

$S - 2bc = 2ab + 2ac$

$S - 2bc = a(2b + 2c)$

$$a = \frac{S - 2bc}{2b + 2c}$$

46. number correct: x

number incorrect: $24 - x$

$25x - 10(24 - x) = 355$

$35x = 595$

$x = 17$

Roger had 17 correct answers

48. time: x

$525 + 0.08x = 700 + 0.05x$

$0.03x = 175$

$x = 5{,}833\frac{1}{3}$

she saves after $5{,}833\frac{1}{3}$ hours

50. number sold: x

$19{,}000 + 0.06(320)x = 21{,}000$

$19.2x = 2{,}000$

$x = 104\frac{1}{6}$

he must sell 105 vacuum cleaners

52. a. price last year: x

$1.06x = 12$

$106x = 1200$

$x = 11.32$

price last year was $11.32

54. a. original holdings: x

current holdings: $0.84x$

$\left(\frac{x}{0.84x}\right)100 - 100 = 19$

He needs a 19% gain

b. No, x cancels out

b.   price two years ago: x

$$(1.06)^2 x = 12$$

$$1.1236x = 12$$

$$x = 10.68$$

price 2 yrs ago was $10.68

56.   batting avg. last 18 wks.: x

$$8(0.385) + 18x = 26(0.350)$$

$$3.08 + 18x = 9.1$$

$$18x = 6.02$$

$$x = 0.334$$

he needs a .334 average

58.   number of juniors: x

number of seniors: 294 - x

$$512x + 526(294 - x) = 294(520)$$

$$512x + 154,644 - 526x = 152,880$$

$$-14x = -1,764$$

$$x = 126$$

126 juniors took the exam

60.   amount of 6% fertilizer: x

amount of 15% : 10 - x

$$0.06x + 0.15(10-x) = 0.08(10)$$

$$6x + 150 - 15x = 80$$

$$-9x = -70$$

$$x = 7\frac{7}{9}$$

she mixes $7\frac{7}{9}$ lb of the 6% fertilizer

and $2\frac{2}{9}$ lb of the 15% fertilizer

62.   Amount invested in stocks: x

$$0.08(3,000) + 0.15x = 0.12(3,000 + x)$$

$$8(3,000) + 15x = 12(3,000 + x)$$

$$24,000 + 15x = 36,000 + 12x$$

$$3x = 12,000$$

$$x = 4,000$$

He should invest 4,000 in stocks

64.   Time to catch them: x

a. $9(x + 2) = 24x$

$$9x + 18 = 24x$$

$$15x = 18$$

$$x = \frac{18}{15} = 1\frac{1}{5} \text{ hr}$$

He needs $1\frac{1}{5}$ hrs to catch up

b. $9\left(3\frac{1}{5}\right)$ or $24\left(1\frac{1}{5}\right) =$

$28\frac{4}{5}$ miles from shore

66.   Time for trip to market: x

Time for return trip: $\frac{7}{12}$ - x

68.   Price of new car: x

$$x + 3(0.085)x = 1200 + 36(220)$$

$$40x = 30\left(\frac{7}{12} - x\right)$$

$$40x = \frac{35}{2} - 30x$$

$$70x = \frac{35}{2}$$

$$x = \frac{1}{4}$$

she lives $40\left(\frac{1}{4}\right)$ or 10 miles from the market

$$1.255x = 9120$$

$$x = \frac{9120}{1.255} = 7266.93$$

They can afford a car that costs $7,266.93

70. Overtime hours: x

$$0.865(8(40) + 8(1.5x)) - 25 = 300$$

$$276.8 + 10.38x = 325$$

$$10.38x = 48.2$$

$$x = 4.644$$

she must work 4.644 or about 4 hours and 39 minutes of overtime per week

Exercise 4.2

2. $4 + \dfrac{x}{5} = \dfrac{5}{3}$

$$15\left(4 + \frac{x}{5}\right) = 15\left(\frac{5}{3}\right)$$

$$60 + 3x = 25$$

$$3x = -35$$

$$x = \frac{-35}{3}$$

the solution is $\dfrac{-35}{3}$

4. $\dfrac{1}{4}x = 2 - \dfrac{1}{3}x$

$$12\left(\frac{1}{4}x\right) = 12\left(2 - \frac{1}{3}x\right)$$

$$3x = 24 - 4x$$

$$7x = 24$$

$$x = \frac{24}{7}$$

the solution is $\dfrac{24}{7}$

6. $\dfrac{x-5}{4} - \dfrac{x-9}{12} = 1$

$$12\left(\frac{x-5}{4} - \frac{x-9}{12}\right) = 12$$

$$3(x-5) - (x-9) = 12$$

$$3x - 15 - x + 9 = 12$$

$$2x = 18$$

$$x = 9$$

the solution is 9

8. $\dfrac{3x}{4} = \dfrac{2}{3} - \dfrac{x-7}{6}$

$$12\left(\frac{3x}{4}\right) = 12\left(\frac{2}{3} - \frac{x-7}{6}\right)$$

10. $\dfrac{5}{x-3} = \dfrac{x+2}{x-3} + 3$

$$(x-3)\left(\frac{5}{x-3}\right) = (x-3)\left(\frac{x+2}{x-3} + 3\right)$$

$9x = 8 - 2x + 14$

$11x = 22$

$x = 2$

the solution is 2

$5 = (x + 2) + 3x - 9$

$4x = 12$

$x = 3$ is extraneous

the equation has no solution

12. $\dfrac{2}{x + 1} + \dfrac{1}{3x + 3} = \dfrac{1}{6}$

$6(x + 1)\left(\dfrac{2}{x + 1} + \dfrac{1}{3(x + 1)}\right) = 6(x + 1)\dfrac{1}{6}$

$12 + 2 = x + 1$

$14 = x + 1$

$x = 13$

the solution is 13

14. $\dfrac{1}{x - 1} + \dfrac{2}{x + 1} = \dfrac{x - 2}{x^2 - 1}$

$(x - 1)(x + 1)\left(\dfrac{1}{x - 1} + \dfrac{2}{x + 1}\right) = (x - 1)(x + 1)\left(\dfrac{x - 2}{(x - 1)(x + 1)}\right)$

$(x + 1) + 2(x - 1) = x - 2$

$x + 1 + 2x - 2 = x - 2$

$2x = -1$

$x = \dfrac{-1}{2}$

the solution is $\dfrac{-1}{2}$

16. $\dfrac{4}{2x - 3} + \dfrac{4x}{4x^2 - 9} = \dfrac{1}{2x + 3}$

$(2x - 3)(2x + 3)\left(\dfrac{4}{2x - 3} + \dfrac{4x}{(2x - 3)(2x + 3)}\right) = (2x - 3)(2x + 3)\left(\dfrac{1}{2x + 3}\right)$

$4(2x + 3) + 4x = 2x - 3$

$8x + 12 + 4x = 2x - 3$

$12x + 12 = 2x - 3$

$10x = -15$

$x = \dfrac{-3}{2}$ is an extraneous solution; the equation has no solution

44

18. $\dfrac{7}{5} = \dfrac{x}{x-2}$

$7(x-2) = 5x$

$7x - 14 = 5x$

$2x = 14$

$x = 7$

the solution is 7

20. $\dfrac{-3}{4} = \dfrac{y-7}{y+14}$

$-3(y+14) = 4(y-7)$

$-3y - 42 = 4y - 28$

$-7y = 14$

$y = -2$

the solution is -2

22. $\dfrac{30}{r} = \dfrac{20}{r-10}$

$30(r-10) = 20r$

$30r - 300 = 20r$

$10r = 300$

$r = 30$

the solution is 30

24. $C = \dfrac{5}{9}(F - 32)$

$9C = 9\left(\dfrac{5}{9}\right)(F - 32)$

$9C = 5(F - 32)$

$9C = 5F - 160$

$5F = 9C + 160$

$F = \dfrac{9}{5}C + 32$

26. $S = \dfrac{n}{2}(a + s)$

$2S = 2\left(\dfrac{n}{2}\right)(a + s)$

$2S = n(a + s)$

$2S = na + ns$

$ns = 2S - na$

$s = \dfrac{2S - na}{n}$

28. $s = vt + \dfrac{1}{2}at^2$

$2s = 2\left(vt + \dfrac{1}{2}at^2\right)$

$2s = 2vt + at^2$

$at^2 = 2s - 2vt$

$a = \dfrac{2s - 2vt}{t^2}$

30. $I^2 = \dfrac{P}{R}$

$RI^2 = R\left(\dfrac{P}{R}\right)$

$R = \dfrac{P}{I^2}$

32. $2.483 = \dfrac{h}{r-6}$

$2.483(r-6) = (r-6)\left(\dfrac{h}{r-6}\right)$

$2.483r - 14.898 = h$

$2.483r = h + 14.898$

$r = \dfrac{h + 14.898}{2.483}$

34. $I = \dfrac{E}{r+R}$

$(r+R)I = (r+R)\left(\dfrac{E}{r+R}\right)$

$rI + RI = E$

$RI = E - rI$

$R = \dfrac{E - rI}{I}$

36. $\dfrac{P}{T} = \dfrac{p}{t}$

$Pt = pT$

$t = \dfrac{pT}{P}$

38. $P = \dfrac{A}{1+rt}$

$(1+rt)P = (1+rt)\left(\dfrac{A}{1+rt}\right)$

$P + Prt = A$

$Prt = A - P$

40. $\dfrac{1}{R} = \dfrac{1}{r} + \dfrac{1}{s}$

$Rrs\left(\dfrac{1}{R}\right) = Rrs\left(\dfrac{1}{r} + \dfrac{1}{s}\right)$

$rs = Rs + Rr$

$rs - Rs = Rr$

$$t = \frac{A - P}{Pr}$$

$$(r - R)s = Rr$$

$$s = \frac{Rr}{r - R}$$

42. $M = \dfrac{ab}{a + b}$

$(a + b)M = (a + b)\left(\dfrac{ab}{a + b}\right)$

$aM + bM = ab$

$aM = ab - bM$

$aM = b(a - M)$

$b = \dfrac{aM}{a - M}$

44. $\dfrac{p}{q} = \dfrac{r}{q + r}$

$p(q + r) = qr$

$pq + pr = qr$

$pq = qr - pr$

$pq = r(q - p)$

$r = \dfrac{pq}{q - p}$

46.   Speed of current: x

$\dfrac{8}{20 - x} = \dfrac{2}{3}\left(\dfrac{18}{20 + x}\right)$

$24(20 - x) = 36(20 - x)$

$480 + 24x = 720 - 36x$

$60x = 240$

$x = 4$

the speed of the current is 4 mph

48.   Distance upriver: x

$\dfrac{x}{6} + \dfrac{x}{10} = 24$

$30\left(\dfrac{x}{6} + \dfrac{x}{10}\right) = 30(24)$

$5x + 3x = 720$

$8x = 720$

$x = 90$

they can go 90 miles upriver

50.   energy cost: x

$\dfrac{70}{83} = \dfrac{x}{1236}$

$86,520 = 83x$

$x = 1,042.41$

the household spends $1,042.41

52.   amount for maintenance: x

amount for new equipment: 320,000-x

$\dfrac{4}{5} = \dfrac{x}{320,000 - x}$

$4(320,000 - x) = 5x$

$1,280,000 - 4x = 5x$

$9x = 1,280,000$

$x = 142,222.22$

$142,222.22 is allocated for maintenance

54.   length: x

$$\frac{8.3}{36} = \frac{11.2}{x}$$

$8.3x = 403.2$

$x = 48.6$

the poster is 48.6 cm. long

56. $$\frac{29.8 \text{ km}}{\text{sec}} = \frac{29.8 \text{ km}}{\frac{1}{60} \text{ min}} = \frac{1788 \text{ km}}{\text{min}} = \frac{1788 \text{ km}}{\frac{1}{60} \text{ hr}} =$$

$$\frac{107{,}280 \text{ km}}{\text{hr}} = \frac{107{,}280(0.621 \text{ mi})}{\text{hr}} = \frac{66{,}620.88 \text{ mi}}{\text{hr}}$$

the velocity is 66,620.88 mph

58.   number of geese: x

$$\frac{x}{30} = \frac{41}{4}$$

$4x = 1230$

$x = 307.5$

approximately 308 geese

60.   height of cliff: x

$$\frac{x}{5.5} = \frac{56}{2}$$

$2x = 308$

$x = 154$

the cliff is 154 feet tall

62.   height of foam cylinder: x

$$\frac{x}{10} = \frac{4 - 2.5}{4}$$

$4x = 15$

$$x = \frac{15}{4}$$

the cylinder is 3.75 inches tall

64.   distance across lake (EC): x

$$\frac{25}{60} = \frac{30}{x}$$

$25x = 60(30)$

$x = 72$

the lake is 72 yards across

66. $$k^2 = \frac{b}{1 + \frac{m}{M}}$$

$$k^2 \left(1 + \frac{m}{M}\right) = b$$

$$k^2 + \frac{k^2 m}{M} = b$$

$$\frac{k^2 m}{M} = b - k^2$$

$$m = \frac{M}{k^2}(b - k^2)$$

68. $$\frac{V^2}{2} = gR^2 \left(\frac{1}{R} - \frac{1}{r}\right)$$

$$\frac{V^2}{2gR^2} = \frac{1}{R} - \frac{1}{r}$$

$$\frac{1}{r} = \frac{1}{R} - \frac{V^2}{2gR^2}$$

$$\frac{1}{r} = \frac{2gR - V^2}{2gR^2}$$

$$r = \frac{2gR^2}{2gR - V^2}$$

70. $V = \dfrac{v_1 + v_2}{1 + \dfrac{v_1 v_2}{c}}$
    $\qquad V\left(1 + \dfrac{v_1 v_2}{c}\right) = v_1 + v_2$

$V + \dfrac{V v_1 v_2}{c} = v_1 + v_2$
$\qquad V - v_2 = v_1 - \dfrac{V v_1 v_2}{c}$

$V - v_2 = v_1\left(1 - \dfrac{V v_2}{c}\right)$
$\qquad v_1 = \dfrac{V - v_2}{1 - \dfrac{V v_2}{c}}$ or $\dfrac{c(V - v_2)}{c - V v_2}$

Exercise 4.3

2. $x + 1 = 0$ or $3x - 1 = 0$

    $x = -1 \qquad x = \dfrac{1}{3}$

    the solutions are -1 and $\dfrac{1}{3}$

4. $x = 0$ or $3x - 7 = 0$

    $x = 0 \qquad x = \dfrac{7}{3}$

    the solutions are 0 and $\dfrac{7}{3}$

6. $2x - 7 = 0$ or $x + 1 = 0$

    $x = \dfrac{7}{2} \qquad x = -1$

    the solutions are $\dfrac{7}{2}$ and -1

8. $3z^2 - 3z = 0$

    $3z(z - 1) = 0$

    $z = 0$ or $z - 1 = 0$

    $z = 0 \qquad z = 1$

    the solutions
    are 0 and 1

10. $3x^2 - 12 = 0$

    $3(x - 2)(x + 2) = 0$

    $x - 2 = 0$ or $x + 2 = 0$

    $x = 2 \qquad x = -2$

    the solutions are
    2 and -2

12. $25x^2 - 4 = 0$

    $(5x - 2)(5x + 2) = 0$

    $5x - 2 = 0$ or $5x + 2 = 0$

    $x = \dfrac{2}{5} \qquad x = \dfrac{-2}{5}$

    the solutions are
    $\dfrac{2}{5}$ and $\dfrac{-2}{5}$

14. $12y^2 - 8y - 15 = 0$

    $(2y - 3)(6y + 5) = 0$

    $2y - 3 = 0$ or $6y + 5 = 0$

    $y = \dfrac{3}{2} \qquad y = \dfrac{-5}{6}$

    the solutions are
    $\dfrac{3}{2}$ and $\dfrac{-5}{6}$

16. $2x^2 - 4x = x + 3$

    $2x^2 - 5x - 3 = 0$

    $(2x + 1)(x - 3) = 0$

    $2x + 1 = 0$ or $x - 3 = 0$

    $x = \dfrac{-1}{2} \qquad x = 3$

    the solutions are $\dfrac{-1}{2}$, 3

18. $5t + 10 = t^2 + 2t$

    $t^2 - 3t - 10 = 0$

20. $x^2 - 2x + 1 = 2x^2 + 3x - 5$

    $x^2 + 5x - 6 = 0$

$(t - 5)(t + 2) = 0$

$t - 5 = 0$ or $t + 2 = 0$

$t = 5$ $\qquad$ $t = -2$

the solutions are 5 and -2

$(x + 6)(x - 1) = 0$

$x + 6 = 0$ or $x - 1 = 0$

$x = -6$ $\qquad$ $x = 1$

the solutions are -6 and 1

22. $6 - 6y = 12 - y^2 - 2y - 1$

$y^2 - 4y - 5 = 0$

$(y - 5)(y + 1) = 0$

$y - 5 = 0$ or $y + 1 = 0$

$y = 5$ $\qquad$ $y = -1$

the solutions are 5 and -1

24. $2x^2 - x - 3 = 3$

$2x^2 - x - 6 = 0$

$(2x + 3)(x - 2) = 0$

$2x + 3 = 0$ or $x - 2 = 0$

$x = \dfrac{-3}{2}$ $\qquad$ $x = 2$

the solutions are $\dfrac{-3}{2}$ and 2

26. $2x - [(x + 2)(x - 3) + 8] = 0$

$2x - (x^2 - x - 6) - 8 = 0$

$0 = x^2 - 3x + 2$

$(x - 2)(x - 1) = 0$

$x - 2 = 0$ or $x - 1 = 0$

$x = 2$ $\qquad$ $x = 1$

the solutions are 2 and 1

28. $3[(x + 2)^2 - 4x] = 15$

$3x^2 + 12 - 15 = 0$

$3(x - 1)(x + 1) = 0$

$x - 1 = 0$ or $x + 1 = 0$

$x = 1$ $\qquad$ $x = -1$

the solutions are 1 and -1

30. $2x - \dfrac{5}{3} = \dfrac{x^2}{3}$

$3\left(2x - \dfrac{5}{3}\right) = 3\left(\dfrac{x^2}{3}\right)$

$x^2 - 6x + 5 = 0$

$(x - 5)(x - 1) = 0$

$x - 5 = 0$ or $x - 1 = 0$

$x = 5$ $\qquad$ $x = 1$

the solutions are 5 and 1

32. $\dfrac{x}{4} - \dfrac{3}{4} = \dfrac{1}{x}$

$4x\left(\dfrac{x}{4} - \dfrac{3}{4}\right) = 4x\left(\dfrac{1}{x}\right)$

$x^2 - 3x - 4 = 0$

$(x - 4)(x + 1) = 0$

$x - 4 = 0$ or $x + 1 = 0$

$x = 4$ $\qquad$ $x = -1$

the solutions are 4 and -1

34. $5 = \dfrac{6}{x^2} - \dfrac{7}{x}$

$x^2(5) = x^2\left(\dfrac{6}{x^2} - \dfrac{7}{x}\right)$

$5x^2 + 7x - 6 = 0$

$(5x - 3)(x + 2) = 0$

$5x - 3 = 0 \text{ or } x + 2 = 0$

$x = \dfrac{3}{5} \qquad x = -2$

the solutions are $\dfrac{3}{5}$ and -2

36. $-3 = \dfrac{-10}{x + 2} + \dfrac{10}{x + 5}$

$-3(x + 2)(x + 5) = (x + 2)(x + 5)\left[\dfrac{-10}{x + 2} + \dfrac{10}{x + 5}\right]$

$-3x^2 - 21x - 30 = -10x - 50 + 10x + 20$

$3x^2 + 21x = 0$

$3x(x + 7) = 0$

$x = 0 \text{ or } x + 7 = 0$

$x = 0 \qquad x = -7$

the solutions are 0 and -7

38. $\dfrac{2}{x^2 - 2x} + \dfrac{1}{2x} = \dfrac{-1}{x^2 + 2x}$

$2x(x - 2)(x + 2)\left(\dfrac{2}{x(x - 2)} + \dfrac{1}{2x}\right) = 2x(x - 2)(x + 2)\left(\dfrac{-1}{x(x + 2)}\right)$

$4x + 8 + x^2 - 4 = -2x + 4$

$x^2 + 6x = 0$

$x(x + 6) = 0$

$x = 0 \text{ or } x + 6 = 0$

$x = 0 \qquad x = -6$

the solution is -6 only (0 is extraneous)

40. $\dfrac{2x + 1}{2x - 3} + \dfrac{x^2 + 3x - 7}{2x^2 - 7x + 6} = \dfrac{x + 1}{x - 2}$

$(2x - 3)(x - 2)\left(\dfrac{2x + 1}{2x - 3} + \dfrac{x^2 + 3x - 7}{(2x - 3)(x - 2)}\right) = (2x - 3)(x - 2)\left(\dfrac{x + 1}{x - 2}\right)$

$2x^2 - 3x - 2 + x^2 + 3x - 7 = 2x^2 - x - 3$

$x^2 + x - 6 = 0$

$(x + 3)(x - 2) = 0$

$x + 3 = 0 \text{ or } x - 2 = 0$

$x = -3 \qquad x = 2$

the solution is -3 only (2 is extraneous)

42. $(x - (-4))(x - 3) = 0$

$x^2 + x - 12 = 0$

44. $(x - 0)(x - 5) = 0$

$x^2 - 5x = 0$

50

46. $3(x - (-\frac{2}{3}))(x - 4) = 0$

$3x^2 - 10x - 8 = 0$

48. $6(x - (-\frac{1}{3}))(x - (-\frac{1}{2})) = 0$

$6x^2 + 5x + 1 = 0$

50. $4x^2 = 9$

$x^2 = \frac{9}{4}$

$x = \frac{3}{2}$ or $x = -\frac{3}{2}$

the solutions are
$\frac{3}{2}$ and $-\frac{3}{2}$

52. $3x^2 = 15$

$x^2 = 5$

$x = \sqrt{5}$ or $x = -\sqrt{5}$

the solutions are
$\sqrt{5}$ and $-\sqrt{5}$

54. $3x^2 + 9 = 0$

$x^2 = -3$

$x = \sqrt{-3}$ or $x = -\sqrt{-3}$

the solutions are
$i\sqrt{3}$ and $-i\sqrt{3}$

56. $\frac{3x^2}{5} = 6$

$x^2 = 10$

$x = \sqrt{10}$ or $x = -\sqrt{10}$

the solutions are
$\sqrt{10}$ and $-\sqrt{10}$

58. $(x + 3)^2 = 4$

$x + 3 = 2$ or $x + 3 = -2$

$x = -1$ or $x = -5$

the solutions are
-1 and -5

60. $(3x + 1)^2 = 25$

$3x + 1 = 5$ or $3x + 1 = -5$

$x = \frac{4}{3}$ or $x = -2$

the solutions are
$\frac{4}{3}$ and -2

62. $(x - 5)^2 = -7$

$x - 5 = \sqrt{-7}$ or $x = -\sqrt{-7}$

$x = 5 + \sqrt{-7}$ or $x = 5 - \sqrt{-7}$

the solutions are $5 + i\sqrt{7}$ and $5 - i\sqrt{7}$

64. $\left(x - \frac{2}{3}\right)^2 = \frac{5}{9}$

$x - \frac{2}{3} = \sqrt{\frac{5}{9}}$ or $x = -\sqrt{\frac{5}{9}}$

$x = \frac{2}{3} + \sqrt{\frac{5}{9}}$ or $x = \frac{2}{3} - \sqrt{\frac{5}{9}}$

the solutions are $\dfrac{2 \pm \sqrt{5}}{3}$

66. $\left(x + \frac{1}{2}\right)^2 = -\frac{1}{16}$

$x + \frac{1}{2} = \sqrt{-\frac{1}{16}}$ or $x + \frac{1}{2} = -\sqrt{-\frac{1}{16}}$

$x = -\frac{1}{2} + \sqrt{-\frac{1}{16}}$ or $x = -\frac{1}{2} - \sqrt{-\frac{1}{16}}$

the solutions are $-\frac{1}{2} + \frac{1}{4}i$ and $-\frac{1}{2} - \frac{1}{4}i$

68. $(5x - 12)^2 = 24$

$5x - 12 = \sqrt{24}$ or $5x - 12 = -\sqrt{24}$

$x = \dfrac{12 + \sqrt{24}}{5}$ or $x = \dfrac{12 - \sqrt{24}}{5}$

the solutions are $\dfrac{12 \pm 2\sqrt{6}}{5}$

70. a. $256 = -16t^2 + 32t + 240$

$16t^2 - 32t + 16 = 0$

$16(t - 1)(t - 1) = 0$

$t - 1 = 0 \text{ or } t - 1 = 0$

it takes 1 second

b. $0 = -16t^2 + 32t + 240$

$16(t - 5)(t + 3) = 0$

$t - 5 = 0 \text{ or } t + 3 = 0$

it takes 5 seconds (since time is never negative)

72. $11\pi = \pi(4)^2 - \pi r^2$

$\pi(r^2 - 5) = 0$

$r^2 = 5$

$r = \sqrt{5} \text{ or } r = -\sqrt{5}$

the radius is $\sqrt{5}$ (since distance cannot be negative)

74. longer side: x

shorter side: x - 4

$x^2 + (x - 4)^2 = 20^2$

$2x^2 - 8x - 384 = 0$

$2(x - 16)(x + 12) = 0$
$x = 16 \text{ or } x = -12$

the rectangle is 16 by 12
(since length is not negative)

76. width: x

length: 3x

$x^2 + (3x)^2 = 60^2$

$10x^2 = 3600$

$x^2 = 360$
$x = \sqrt{360} \text{ or } -\sqrt{360}$

the rectangle is $6\sqrt{10}$ feet wide
and $18\sqrt{10}$ feet long

78. width of border: x

$(2x + 15)(x + 12) - 12(15) = 135$

$2x^2 + 39x - 135 = 0$

$(2x + 45)(x - 3) = 0$

$x = -22.5 \text{ or } x = 3$

the border is 3 feet wide

80. side next to river: x

remaining side: 360 - 2x

$x(360 - 2x) = 16,000$

$2x^2 - 360x + 16,000 = 0$

$2(x - 100)(x - 80) = 0$

$x = 100 \text{ or } x = 80$

the pasture is 100 by 160 or 80 by 200

82. amount turned up: x

$6x(1 - 2x) = \dfrac{3}{4}$

$24x - 48x^2 = 3$

84. number of $0.50 increases: x

$(80 - 4x)(8 + 0.5x) = 648$

$640 + 8x - 2x^2 = 648$

$48x^2 - 24x + 3 = 0$

$3(4x - 1)(4x - 1) = 0$

$x = \dfrac{1}{4}$

sides turned up $\dfrac{1}{4}$ inch

86. $\dfrac{300}{x} = 5x - 55$

$5x^2 - 55x - 300 = 0$

$5(x + 4)(x - 15) = 0$

$x = -4$ or $x = 15$

she sells all she buys at $15

(she sells 20 dozen per week)

90. $(2a + x)^2 - b^2 = 0$

$(2a + x)^2 = b^2$

$2a + x = b$ or $2a + x = -b$

$x = b - 2a$ or $x = -b - 2a$

94. $x^2 - \dfrac{a^2 + b^2}{ab}x + 1 = 0$

$\left(x - \dfrac{a}{b}\right)\left(x - \dfrac{b}{a}\right) = 0$

$x = \dfrac{a}{b}$ or $x = \dfrac{b}{a}$

98. if the solutions are r and s,

then $(x - r)(x - s) = 0$, so

$x^2 - (r + s)x + rs = 0$ or

$ax^2 - a(r + s)x + ars = 0$.

$2x^2 - 8x + 8 = 0$

$2(x - 2)(x - 2) = 0$

$x = 2$

increase the price by $1.00 to $9.00

88. speed of current: x

$\dfrac{90}{20 - x} = \dfrac{75}{20 + x} + 3$

$(400 - x^2)\left(\dfrac{90}{20 - x}\right) = (400 - x^2)\left(\dfrac{75}{20 + x} + 3\right)$

$1800 + 90x = 1500 - 75x + 1200 - 3x^2$

$3x^2 + 165x - 900 = 0$

$3(x + 60)(x - 5) = 0$

$x = -60$ or $x = 5$

the current is 5 mph

92. $a^2x^2 + 2a(b + c)x + (b + c)^2 = 0$

$(ax + (b + c))(ax + (b + c)) = 0$

$ax + (b + c) = 0$ or $ax + (b + c) = 0$

$x = -\dfrac{b + c}{a}$

96. $36a + 96 - 12 = 0$

$a = -\dfrac{7}{3}$

$-\dfrac{7}{3}x^2 + 16x - 12 = 0$

$7x^2 - 48x + 36 = 0$

$(7x - 6)(x - 6) = 0$

$7x - 6 = 0$ or $x - 6 = 0$

the other solution is $\dfrac{6}{7}$

Then $ax^2 + bx + c =$

$ax^2 - a(r + s)x + ars$, so

$-a(r + s) = b$ and $ars = c$.

Thus $r + s = -\dfrac{b}{a}$ and $rs = \dfrac{c}{a}$.

Exercise 4.4

2. $x^2 - 14x + 49 = (x - 7)^2$

4. $x^2 + 3x + \dfrac{9}{4} = \left(x + \dfrac{3}{2}\right)^2$

6. $x^2 - \dfrac{5}{2}x + \dfrac{25}{16} = \left(x - \dfrac{5}{4}\right)^2$

8. $x^2 + \dfrac{2}{3}x + \dfrac{1}{9} = \left(x + \dfrac{1}{3}\right)^2$

10. $x^2 + 4x + 4 = 0$

$(x + 2)^2 = 0$

$x + 2 = 0$

$x = -2$

the solution is -2

12. $x^2 - x - 20 = 0$

$x^2 - x + \dfrac{1}{4} = 20 + \dfrac{1}{4}$

$\left(x - \dfrac{1}{2}\right)^2 = \dfrac{81}{4}$

$x - \dfrac{1}{2} = \dfrac{9}{2}$ or $x - \dfrac{1}{2} = -\dfrac{9}{2}$

$x = 5$ or $x = -4$

the solutions are 5 and -4

14. $x^2 = 5 - 5x$

$x^2 + 5x + \dfrac{25}{4} = 5 + \dfrac{25}{4}$

$\left(x + \dfrac{5}{2}\right)^2 = \dfrac{45}{4}$

$x + \dfrac{5}{2} = \sqrt{\dfrac{45}{4}}$ or $x + \dfrac{5}{2} = -\sqrt{\dfrac{45}{4}}$

$x = -\dfrac{5}{2} + \sqrt{\dfrac{45}{4}}$ $x = -\dfrac{5}{2} - \sqrt{\dfrac{45}{4}}$

the solutions are $\dfrac{-5 + 3\sqrt{5}}{2}$ and $\dfrac{-5 - 3\sqrt{5}}{2}$

16. $3x^2 + x - 4 = 0$

$x^2 + \dfrac{1}{3}x + \dfrac{1}{36} = \dfrac{4}{3} + \dfrac{1}{36}$

$\left(x + \dfrac{1}{6}\right)^2 = \dfrac{49}{36}$

$x + \dfrac{1}{6} = \dfrac{7}{6}$ or $x + \dfrac{1}{6} = -\dfrac{7}{6}$

$x = -\dfrac{1}{6} + \dfrac{7}{6}$ $x = -\dfrac{1}{6} - \dfrac{7}{6}$

the solutions are 1 and $-\dfrac{4}{3}$

18. $4x^2 - 3 = 2x$

$$x^2 - \frac{1}{2}x + \frac{1}{16} = \frac{3}{4} + \frac{1}{16}$$

$$\left(x - \frac{1}{4}\right)^2 = \frac{13}{16}$$

$$x - \frac{1}{4} = \sqrt{\frac{13}{16}} \text{ or } x - \frac{1}{4} = -\sqrt{\frac{13}{16}}$$

$$x = \frac{1}{4} + \sqrt{\frac{13}{16}} \quad x = \frac{1}{4} - \sqrt{\frac{13}{16}}$$

the solutions are $\dfrac{1 \pm \sqrt{13}}{4}$

20. $3x^2 + x = -4$

$$x^2 + \frac{1}{3}x + \frac{1}{36} = -\frac{4}{3} + \frac{1}{36}$$

$$\left(x + \frac{1}{6}\right)^2 = -\frac{47}{36}$$

$$x + \frac{1}{6} = \sqrt{-\frac{47}{36}} \text{ or } x + \frac{1}{6} = -\sqrt{-\frac{47}{36}}$$

$$x = -\frac{1}{6} + \sqrt{-\frac{47}{36}} \quad x = -\frac{1}{6} - \sqrt{-\frac{47}{36}}$$

the solutions are $\dfrac{-1 \pm i\sqrt{47}}{6}$

22. $x^2 - 4x + 4 = 0$

$$x = \frac{-(-4) \pm \sqrt{(-4)^2 - 4(1)(4)}}{2(1)}$$

$$x = \frac{4 \pm \sqrt{0}}{2}$$

the solution is 2

24. $y^2 - 5y = 6$

$$y^2 - 5y - 6 = 0$$

$$y = \frac{-(-5) \pm \sqrt{(-5)^2 - 4(1)(-6)}}{2(1)}$$

$$y = \frac{5 \pm \sqrt{25 + 24}}{2} = \frac{5 \pm 7}{2}$$

the solutions are 6 and -1

26. $2z^2 = 7z - 6$

$$2z^2 - 7z + 6 = 0$$

$$z = \frac{-(-7) \pm \sqrt{(-7)^2 - 4(2)(6)}}{2(2)}$$

$$z = \frac{7 \pm \sqrt{49 - 48}}{4} = \frac{7 \pm 1}{4}$$

the solutions are 2 and $\dfrac{3}{2}$

28. $0 = x^2 - \dfrac{1}{2}x + \dfrac{1}{2}$

$$2x^2 - x + 1 = 0$$

$$x = \frac{-(-1) \pm \sqrt{(-1)^2 - 4(2)(1)}}{2(2)}$$

$$x = \frac{1 \pm \sqrt{1 - 8}}{4} = \frac{1 \pm \sqrt{-7}}{4}$$

the solutions are $\dfrac{1 \pm i\sqrt{7}}{4}$

30. $13z + 5 = 6z^2$

$$6z^2 - 13z - 5 = 0$$

32. $y^2 + 3y = 0$

$$y = \frac{-3 \pm \sqrt{3^2 - 4(1)(0)}}{2(1)}$$

$$z = \frac{-(-13) \pm \sqrt{(-13)^2 - 4(6)(-5)}}{2(6)}$$

$$z = \frac{13 \pm \sqrt{169 + 120}}{12} = \frac{13 \pm 17}{12}$$

the solutions are $\frac{5}{2}$ and $-\frac{1}{3}$

$$y = \frac{-3 \pm \sqrt{9}}{2} = \frac{-3 \pm 3}{2}$$

the solutions are -3 and 0

34. $2z^2 + 1 = 0$

$$z = \frac{-0 \pm \sqrt{0^2 - 4(2)(1)}}{2(2)}$$

$$z = \frac{\sqrt{-8}}{4} = \frac{2i\sqrt{2}}{4}$$

the solutions are $\pm \frac{i\sqrt{2}}{2}$

36. $x^2 + 2x = -5$

$x^2 + 2x + 5 = 0$

$$x = \frac{-2 \pm \sqrt{2^2 - 4(1)(5)}}{2(1)}$$

$$x = \frac{-2 \pm \sqrt{4 - 20}}{2} = \frac{-2 \pm \sqrt{-16}}{2}$$

the solutions are $-1 \pm 2i$

38. $2z = \frac{3}{z - 2}$

$2z^2 - 4z = 3$

$2z^2 - 4z - 3 = 0$

$$z = \frac{-(-4) \pm \sqrt{(-4)^2 - 4(2)(-3)}}{2(2)}$$

$$z = \frac{4 \pm \sqrt{16 + 24}}{4} = \frac{4 \pm \sqrt{40}}{4}$$

the solutions are $\frac{2 \pm \sqrt{10}}{2}$

40. $2x = \frac{x + 1}{x - 1}$

$2x^2 - 2x = x + 1$

$2x^2 - 3x - 1 = 0$

$$x = \frac{-(-3) \pm \sqrt{(-3)^2 - 4(2)(-1)}}{2(2)}$$

$$x = \frac{3 \pm \sqrt{9 + 8}}{4} = \frac{3 \pm \sqrt{17}}{4}$$

the solutions are $\frac{3 \pm \sqrt{17}}{4}$

42. $3z^2 + 2z + 2 = z$

$3z^2 + z + 2 = 0$

$$z = \frac{-1 \pm \sqrt{1^2 - 4(3)(2)}}{2(3)}$$

$$z = \frac{-1 \pm \sqrt{1 - 24}}{6} = \frac{-1 \pm \sqrt{-23}}{6}$$

the solutions are $\frac{-1 \pm i\sqrt{23}}{6}$

44. $y = \frac{-1}{y - 1}$

$y^2 - y + 1 = 0$

$$y = \frac{-(-1) \pm \sqrt{(-1)^2 - 4(1)(1)}}{2(1)}$$

$$y = \frac{1 \pm \sqrt{1 - 4}}{2} = \frac{1 \pm \sqrt{-3}}{2}$$

the solutions are $\frac{1 \pm i\sqrt{3}}{2}$

46. $\dfrac{3x + 1}{x^2 + x + 5} = 1$

$3x + 1 = x^2 + x + 5$

$x^2 - 2x + 4 = 0$

$x = \dfrac{-(-2) \pm \sqrt{(-2)^2 - 4(1)(4)}}{2(1)}$

$x = \dfrac{2 \pm \sqrt{4 - 16}}{2} = \dfrac{2 \pm \sqrt{-12}}{2}$

the solutions are $1 \pm i\sqrt{3}$

48. $\dfrac{3}{2x + 1} - \dfrac{2x - 3}{x} = 0$

$x(2x + 1)\left(\dfrac{3}{2x + 1} - \dfrac{2x - 3}{x}\right) = 0x(2x + 1)$

$3x - (4x^2 - 4x - 3) = 0$

$4x^2 - 7x - 3 = 0$

$x = \dfrac{-(-7) \pm \sqrt{(-7)^2 - 4(4)(-3)}}{2(4)}$

$x = \dfrac{7 \pm \sqrt{49 + 48}}{8} = \dfrac{7 \pm \sqrt{97}}{8}$

the solutions are $\dfrac{7 \pm \sqrt{97}}{8}$

50. $\dfrac{3x - 1}{3x + 1} + \dfrac{2x}{2x + 1} = 1$

$(3x + 1)(2x + 1)\left(\dfrac{3x - 1}{3x + 1} + \dfrac{2x}{2x + 1}\right) = 1(3x + 1)(2x + 1)$

$6x^2 + x - 1 + 6x^2 + 2x = 6x^2 + 5x + 1$

$6x^2 - 2x - 2 = 0$

$x = \dfrac{-(-2) \pm \sqrt{(-2)^2 - 4(6)(-2)}}{2(6)} = \dfrac{2 \pm \sqrt{4 + 48}}{12} = \dfrac{2 \pm \sqrt{52}}{12}$

the solutions are $\dfrac{1 \pm \sqrt{13}}{6}$

52. $K = \dfrac{1}{2}mv^2$

$v^2 = \dfrac{2K}{m}$

$v = \pm\sqrt{\dfrac{2K}{m}}$

54. $A = \dfrac{\sqrt{3}}{4}s^2$

$s^2 = \dfrac{4A}{\sqrt{3}}$

$s = \pm\dfrac{2\sqrt{A}}{\sqrt[4]{3}}$

56. $V = \pi(r - 2)^2 h$

$(r - 2)^2 = \dfrac{V}{\pi h}$

$r - 2 = \pm\sqrt{\dfrac{V}{\pi h}}$

$r = 2 \pm\sqrt{\dfrac{V}{\pi h}}$

58. $F = \dfrac{kq_1q_2}{r^2}$

$Fr^2 = kq_1q_2$

$r^2 = \dfrac{kq_1q_2}{F}$

$r = \pm\sqrt{\dfrac{kq_1q_2}{F}}$

60. $V = 2(s^2 + t^2)w$

$V = 2ws^2 + 2wt^2$

$2wt^2 = V - 2ws^2$

$t^2 = \dfrac{V - 2ws^2}{2w}$

$t = \pm\sqrt{\dfrac{V - 2ws^2}{2w}}$

62. $2a^2 + 3b^2 = c^2$

$2a^2 = c^2 - 3b^2$

$a^2 = \dfrac{c^2 - 3b^2}{2}$

$a = \pm\sqrt{\dfrac{c^2 - 3b^2}{2}}$

64. $\dfrac{x^2}{5} - \dfrac{y^2}{6} = 1$

$30\left(\dfrac{x^2}{5} - \dfrac{y^2}{6}\right) = 30$

$6x^2 - 5y^2 = 30$

$x^2 = \dfrac{30 + 5y^2}{6}$

$x = \pm\sqrt{\dfrac{5y^2 + 30}{6}}$

66. $h = \dfrac{1}{2}gt^2 + dl$

$\dfrac{1}{2}gt^2 = h - dl$

$t^2 = \dfrac{2h - 2dl}{g}$

$t = \pm\sqrt{\dfrac{2h - 2dl}{g}}$

68. $R^2 = \dfrac{E^2 - I^2(\omega L)^2}{I^2}$

$R^2 = \dfrac{E^2}{I^2} - (\omega L)^2$

$\dfrac{E^2}{I^2} = R^2 + (\omega L)^2$

$\dfrac{E^2}{R^2 + (\omega L)^2} = I^2$

$I = \pm\dfrac{E}{\sqrt{R^2 + (\omega L)^2}}$

70. $m = \dfrac{4A^2}{a(1 - e^2)}$

$a - ae^2 = \dfrac{4A^2}{m}$

$ae^2 = a - \dfrac{4A^2}{m}$

$e^2 = 1 - \dfrac{4A^2}{ma}$

$e = \pm\sqrt{1 - \dfrac{4A^2}{ma}}$

72. $A = \pi r^2 + \pi rs$

$\pi r^2 + \pi sr - A = 0$

$r = \dfrac{-\pi s \pm \sqrt{s^2\pi^2 - 4(\pi)(-A)}}{2\pi}$

$r = \dfrac{-\pi s \pm \sqrt{s^2\pi^2 + 4A\pi}}{2\pi}$

74. $h = 6t - 3t^2$

$3t^2 - 6t + h = 0$

$t = \dfrac{-(-6) \pm \sqrt{(-6)^2 - 4(3)h}}{2(3)}$

$t = \dfrac{3 \pm \sqrt{9 - 3h}}{3}$

76. $s = vt - \dfrac{1}{2}at^2$

$at^2 - 2vt + 2s = 0$

78. $S = \dfrac{n^2 + n}{2}$

$n^2 + n - 2S = 0$

80. $y^2 - 3xy + x^2 = 3$

$x^2 - 3yx + (y^2 - 3) = 0$

$$t = \frac{2v \pm \sqrt{4v^2 - 8as}}{2a}$$

$$n = \frac{-1 \pm \sqrt{1 - 4(1)(-2S)}}{2(1)}$$

$$x = \frac{3y \pm \sqrt{9y^2 - 4y^2 + 12}}{2}$$

$$t = \frac{v \pm \sqrt{v^2 - 2as}}{a}$$

$$n = \frac{-1 \pm \sqrt{1 + 8s}}{2}$$

$$x = \frac{3y \pm \sqrt{5y^2 + 12}}{2}$$

82. $100 = \frac{s^2}{12} + \frac{s}{2}$

$s^2 + 6s - 1200 = 0$

$s = \frac{-6 \pm \sqrt{4836}}{2}$

the car must have been traveling faster than $-3 + \sqrt{1209}$ mph, or about 31.77 mph

84. a. $10 = -9.8t^2 + 8t + 10$

$9.8t^2 - 8t = 0$

$t(9.8t - 8) = 0$

$t = 0$ or $t = 0.816$

he passes the board after 0.816 sec

86. annual inflation rate: x

$1200(1 + x)^2 = 1400$

$(1 + x)^2 = \frac{7}{6}$

$1 + x = \sqrt{\frac{7}{6}}$

$x = \sqrt{\frac{7}{6}} - 1$

the inflation rate is 0.08 or 8%

b. $0 = -9.8t^2 + 8t + 10$

$9.8t^2 - 8t - 10 = 0$

$t = \frac{8 + \sqrt{456}}{18.6}$

he hits the water after 1.58 sec

88. radius of semicircle: x

width of window: 2x

length of window: 2x + 2

$120 = 2x(2x + 2) + \frac{1}{2}\pi x^2$

$\pi x^2 + 8x^2 + 8x - 240 = 0$

$x = \frac{-8 \pm \sqrt{64 + 4(240)(\pi + 8)}}{2(\pi + 8)}$

$x = \frac{-4 \pm \sqrt{1936 + 240\pi}}{\pi + 8}$

the window is approximately 8.59 feet wide and 10.59 feet high

90. a. distance visible: x

$x^2 + 6370^2 = (6370 + 10)^2$

$x^2 = 40,704,400 - 40,576,900$

$x^2 = 127,500$

$x = \sqrt{127,500} = 50\sqrt{51}$

they can see about 357 km

b. height: x

$10^2 + 6370^2 = (6370 + x)^2$

$$6370 + x = \sqrt{40{,}577{,}000}$$

$$x = 0.00785 \text{ km} = 7.85 \text{ m}$$

they must be 7.85 m high

92. second solution is $3 + \sqrt{2}$

$$(x - (3 + \sqrt{2}\,))(x - (3 - \sqrt{2}\,)) = 0$$

$$x^2 - 6x + 7 = 0$$

94. second solution is 5 - i

$$(x - (5 - i))(x - (5 + i)) = 0$$

$$x^2 - 10x + 26 = 0$$

96. $ax^2 + bx + c = 0$

$$x^2 + \frac{b}{a}x = -\frac{c}{a}$$

$$x^2 + \frac{b}{a}x + \left(\frac{b}{2a}\right)^2 = -\frac{c}{a} + \left(\frac{b}{2a}\right)^2$$

$$\left(x + \frac{b}{2a}\right)^2 = \frac{-4ac}{4a^2} + \frac{b^2}{4a^2}$$

$$x + \frac{b}{2a} = \pm\sqrt{\frac{b^2 - 4ac}{4a^2}}$$

$$x = \frac{-b}{2a} \pm \frac{\sqrt{b^2 - 4ac}}{2a} = \frac{-b \pm \sqrt{b^2 - 4ac}}{2a}$$

this is the quadratic formula

98. $D = 23^2 - 4(4)(-19)$

$D = 833 > 0$

the equation has two distinct real solutions

100. $D = 1254^2 - 4(121)(3249)$

$D = 1{,}572{,}516 - 1{,}572{,}516$

$D = 0$

the equation has one real root (of multiplicity two)

102. $D = (-0.05)^2 - 4(0.03)(0.12) = -0.0119$

the equation has two complex (conjugate) solutions

Exercise 4.5

2. $\sqrt{x} - 4 = 1$

$\sqrt{x} = 5$

$(\sqrt{x})^2 = 5^2$

$x = 25$

the solution is 25

4. $\sqrt{y - 3} = 5$

$(\sqrt{y - 3})^2 = 5^2$

$y - 3 = 25$

$y = 28$

the solution is 28

6. $2z - 3 = \sqrt{7z - 3}$

$(2z - 3)^2 = (\sqrt{7z - 3})^2$

$4z^2 - 12z + 9 = 7z - 3$

$4z^2 - 19z + 12 = 0$

$(4z - 3)(z - 4) = 0$

$z = \frac{3}{4}$ or $z = 4$

the solution is 4 $\left(\frac{3}{4} \text{ is extraneous}\right)$

8. $4x + 5 = \sqrt{3x + 4}$

$(4x + 5)^2 = \left(\sqrt{3x + 4}\right)^2$

$16x^2 + 40x + 25 = 3x + 4$

$16x^2 + 37x + 21 = 0$

$(16x + 21)(x + 1) = 0$

$x = \dfrac{-21}{16}$ or $x = -1$

the solution is -1 $\left(\dfrac{-21}{16}\text{ is extraneous}\right)$

10. $4\sqrt{x - 4} = x$

$\left(4\sqrt{x - 4}\right)^2 = x^2$

$16x - 64 = x^2$

$x^2 - 16x + 64 = 0$

$(x - 8)^2 = 0$

$x = 8$

the solution is 8

12. $\sqrt{4y + 1} = \sqrt{6y - 3}$

$\left(\sqrt{4y + 1}\right)^2 = \left(\sqrt{6y - 3}\right)^2$

$4y + 1 = 6y - 3$

$2y = 4$

$y = 2$

the solution is 2

14. $\sqrt{x}\,\sqrt{x - 5} = 6$

$\left(\sqrt{x}\,\sqrt{x - 5}\right)^2 = 6^2$

$x^2 - 5x = 36$

$x^2 - 5x - 36 = 0$

$(x - 9)(x + 4) = 0$

$x = 9$ or $x = -4$

the solution is 9 (-4 is extraneous)

16. $4\sqrt{y} + \sqrt{1 + 16y} = 5$

$\left(\sqrt{1 + 16y}\right)^2 = \left(5 - 4\sqrt{y}\right)^2$

$1 + 16y = 25 - 40\sqrt{y} + 16y$

$\left(5\sqrt{y}\right)^2 = 3^2$

$25y = 9$

$y = \dfrac{9}{25}$

the solution is $\dfrac{9}{25}$

18. $\sqrt{4x + 17} = 4 - \sqrt{x + 1}$

$\left(\sqrt{4x + 17}\right)^2 = \left(4 - \sqrt{x + 1}\right)^2$

$4x + 17 = 16 - 8\sqrt{x + 1} + x + 1$

$(3x)^2 = \left(-8\sqrt{x + 1}\right)^2$

$9x^2 = 64x + 64$

$(9x + 8)(x - 8) = 0$

$x = \dfrac{-8}{9}$ or $x = 8$

the solution is $\dfrac{-8}{9}$ (8 is extraneous)

20. $(y + 7)^{1/2} + (y + 4)^{1/2} = 3$

$\left[(y + 7)^{1/2}\right]^2 = \left[3 - (y + 4)^{1/2}\right]^2$

$y + 7 = 9 - 6(y + 4)^{1/2} + y + 4$

$\left[(y + 4)^{1/2}\right]^2 = 1^2$

$y + 4 = 1$

$y = -3$

the solution is -3

22. $(z - 3)^{1/2} + (z + 5)^{1/2} = 4$

$\left[(z - 3)^{1/2}\right]^2 = \left[4 - (z + 5)^{1/2}\right]^2$

$z - 3 = 16 - 8(z + 5)^{1/2} + z + 5$

$\left[(z + 5)^{1/2}\right]^2 = 3^2$

$z + 5 = 9$

$z = 4$

the solution is 4

24. $\sqrt[3]{x} = -4$

$\left(\sqrt[3]{x}\right)^3 = (-4)^3$

$x = -64$

the solution is -64

26. $\sqrt[4]{x - 1} = 3$

$\left(\sqrt[4]{x - 1}\right)^4 = 3^4$

$x - 1 = 81$

$x = 82$

the solution is 82

28. $x^{3/4} + 3 = 11$

$\left(x^{3/4}\right)^4 = 8^4$

$x^3 = 4096$

$x = 16$

the solution is 16

30. $(6x - 2)^{5/3} = -32$

$\left[(6x - 2)^{5/3}\right]^3 = (-32)^3$

$(6x - 2)^5 = -32,768$

$6x - 2 = -8$

$x = -1$

the solution is -1

32. $x^{-3/2} = 8$

$\left(x^{-3/2}\right)^2 = 8^2$

$x^{-3} = 64$

$x^3 = \dfrac{1}{64}$

$x = \dfrac{1}{4}$

the solution is $\dfrac{1}{4}$

34. $(5x + 2)^{-1/3} = \dfrac{1}{4}$

$\left[(5x + 2)^{-1/3}\right]^3 = \left(\dfrac{1}{4}\right)^3$

$(5x + 2)^{-1} = \dfrac{1}{64}$

$5x + 2 = 64$

$x = \dfrac{62}{5}$

the solution is $\dfrac{62}{5}$

36. $\sqrt[4]{x^3 - 7} = 3$

$\left(\sqrt[4]{x^3 - 7}\right)^4 = 3^4$

38. $\sqrt[3]{2x^2 - 11x} = 6$

$\left(\sqrt[3]{2x^2 - 11x}\right)^3 = 6^3$

40. $T = 2\pi \sqrt{\dfrac{m}{k}}$

$\left(\sqrt{\dfrac{m}{k}}\right)^2 = \left(\dfrac{T}{2\pi}\right)^2$

$$x^3 - 7 = 81 \qquad\qquad 2x^2 - 11x = 216 \qquad\qquad \frac{m}{k} = \frac{T^2}{4\pi^2}$$

$$x^3 = 88 \qquad\qquad (2x - 27)(x + 8) = 0 \qquad\qquad m = \frac{kT^2}{4\pi^2}$$

$$x = \sqrt[3]{88} \qquad\qquad x = \frac{27}{2} \text{ or } x = -8$$

the solution is $\sqrt[3]{88}$ $\qquad\qquad$ the solutions are $\frac{27}{2}$ and -8

42. $\quad S = r\sqrt{\dfrac{2g}{r+h}}$ $\qquad$ 44. $\quad d = \sqrt[3]{\dfrac{16Mr^2}{m}}$ $\qquad$ 46. $\quad T = \sqrt[4]{\dfrac{E}{SA}}$

$$S^2 = \left[r\sqrt{\frac{2g}{r+h}} \right]^2 \qquad d^3 = \left[\sqrt[3]{\frac{16Mr^2}{m}} \right]^3 \qquad T^4 = \left[\sqrt[4]{\frac{E}{sA}} \right]^4$$

$$S^2 = \frac{2r^2 g}{r+h} \qquad\qquad d^3 = \frac{16Mr^2}{m} \qquad\qquad T^4 = \frac{E}{sA}$$

$$r + h = \frac{2r^2 g}{S^2} \qquad\qquad M = \frac{md^3}{16r^2} \qquad\qquad sA = \frac{E}{T^4}$$

$$h = \frac{2r^2 g}{S^2} - r \qquad\qquad\qquad\qquad\qquad\qquad A = \frac{E}{sT^4}$$

48. $\quad c = \sqrt{a^2 - b^2}$ $\qquad\qquad$ 50. $\quad A = B - C\sqrt{D - E^2}$

$$c^2 = \left(\sqrt{a^2 - b^2} \right)^2 \qquad\qquad \left(C\sqrt{D - E^2} \right)^2 = (B - A)^2$$
$$c^2 = a^2 - b^2 \qquad\qquad\qquad\qquad C^2 D - C^2 E^2 = (B - A)^2$$

$$b^2 = a^2 - c^2 \qquad\qquad\qquad\qquad C^2 E^2 = C^2 D - (B - A)^2$$

$$b = \pm\sqrt{a^2 - c^2} \qquad\qquad\qquad\qquad E^2 = \frac{C^2 D - (B - A)^2}{C^2}$$

$$E = \pm\frac{\sqrt{C^2 D - (B - A)^2}}{C}$$

52. $\quad f = \dfrac{30}{\pi}\sqrt{\dfrac{g}{l^2 - r^2}}$ $\qquad\qquad$ 54. $\quad k = \dfrac{2\pi}{p}\sqrt{\dfrac{(a + R)^3}{1 + m}}$

$$f^2 = \left[\frac{30}{\pi}\sqrt{\frac{g}{l^2 - r^2}} \right]^2 \qquad\qquad k^2 = \left[\frac{2\pi}{p}\sqrt{\frac{(a + R)^3}{1 + m}} \right]^2$$

$$f^2 = \frac{900g}{\pi^2(l^2 - r^2)}$$

$$l^2 - r^2 = \frac{900g}{\pi^2 f^2}$$

$$l^2 = r^2 + \frac{900g}{\pi^2 f^2}$$

$$l = \pm \sqrt{r^2 + \frac{900g}{\pi^2 f^2}}$$

$$k^2 = \frac{4\pi^2(a + R)^3}{p^2(1 + m)}$$

$$(a + R)^3 = \frac{p^2 k^2(1 + m)}{4\pi^2}$$

$$a + R = \sqrt[3]{\frac{p^2 k^2(1 + m)}{4\pi^2}}$$

$$a = \sqrt[3]{\frac{p^2 k^2(1 + m)}{4\pi^2}} - R$$

56.

$$V = \sqrt{k\left(\frac{2}{r} - \frac{1}{a}\right)}$$

$$V^2 = \left[\sqrt{k\left(\frac{2}{r} - \frac{1}{a}\right)}\right]^2$$

$$k\left(\frac{2}{r} - \frac{1}{a}\right) = V^2$$

$$\frac{2}{r} = \frac{1}{a} + \frac{V^2}{k} = \frac{k + aV^2}{ak}$$

$$\frac{r}{2} = \frac{ak}{k + aV^2}$$

$$r = \frac{2ak}{k + aV^2}$$

58. $y^4 - 6y^2 + 5 = 0$

make the substitution $y^2 = u$

$u^2 - 6u + 5 = 0$

$(u - 5)(u - 1) = 0$

$u = 5$ or $u = 1$

$y^2 = 5$ or $y^2 = 1$

the solutions are $\pm \sqrt{5}$ and ± 1

60. $3z^6 - 7z^3 = 6$

$3z^6 - 7z^3 - 6 = 0$

make the substitution $z^3 = u$

$3u^2 - 7u - 6 = 0$

$(3u + 2)(u - 3) = 0$

$u = -\frac{2}{3}$ or $u = 3$

$z^3 = -\frac{2}{3}$ or $z^3 = 3$

the solutions are $-\sqrt[3]{\frac{2}{3}}$ and $\sqrt[3]{3}$

62. $x + 3\sqrt{x} = 10$

make the substitution $\sqrt{x} = u$

$u^2 + 3u - 10 = 0$

$(u + 5)(u - 2) = 0$

$u = -5$ or $u = 2$

$\sqrt{x} = -5$ or $\sqrt{x} = 2$

the solution is 4 (25 is extraneous)

64

64. $\sqrt[3]{x^2} - 12\sqrt[3]{x} + 20 = 0$

make the substitution $\sqrt[3]{x} = u$

$u^2 - 12u + 20 = 0$

$(u - 10)(u - 2) = 0$

$u = 10$ or $u = 2$

$\sqrt[3]{x} = 10$ or $\sqrt[3]{x} = 2$

the solutions are 1000 and 8

66. $\sqrt{x - 6} + 3\sqrt[4]{x - 6} = 18$

make the substitution $\sqrt[4]{x - 6} = u$

$u^2 + 3u - 18 = 0$

$(u + 6)(u - 3) = 0$

$u = -6$ or $u = 3$

$\sqrt[4]{x - 6} = -6$ or $\sqrt[4]{x - 6} = 3$

$x - 6 = 1296$ or $x - 6 = 81$

the solution is 87 (1302 is extraneous)

68. $27z^3 + 26z^{3/2} - 1 = 0$

make the substitution $z^{3/2} = u$

$27u^2 + 26u - 1 = 0$

$(27u - 1)(u + 1) = 0$

$u = \dfrac{1}{27}$ or $u = -1$

$z^{3/2} = \dfrac{1}{27}$ or $z^{3/2} = -1$

$z = \dfrac{1}{9}$ or $z = 1$

the solution is $\dfrac{1}{9}$ (1 is extraneous)

70. $2y^{2/3} + 5y^{1/3} = 3$

make the substitution $y^{1/3} = u$

$2u^2 + 5u - 3 = 0$

$(2u - 1)(u + 3) = 0$

$u = \dfrac{1}{2}$ or $u = -3$

$y^{1/3} = \dfrac{1}{2}$ or $y^{1/3} = -3$

$y = \dfrac{1}{8}$ or $y = -27$

the solutions are $\dfrac{1}{8}$ and -27

72. $8x^{1/2} + 7x^{1/4} = 1$

make the substitution $x^{1/4} = u$

$8u^2 + 7u - 1 = 0$

$(8u - 1)(u + 1) = 0$

$u = \dfrac{1}{8}$ or $u = -1$

$x^{1/4} = \dfrac{1}{8}$ or $x^{1/4} = -1$

the solution is $\dfrac{1}{4096}$ (1 is extraneous)

74. $z^{-2} + 9z^{-1} - 10 = 0$

make the substitution $z^{-1} = u$

$u^2 + 9u - 10 = 0$

$(u - 1)(u + 10) = 0$

$u = 1$ or $u = -10$

$z^{-1} = 1$ or $z^{-1} = -10$

the solutions are 1 and $\dfrac{-1}{10}$

76. $(x - 2)^{1/2} - 11(x - 2)^{1/4} + 18 = 0$

make the substitution $(x - 2)^{1/4} = u$

$u^2 - 11u + 18 = 0$

$(u - 9)(u - 2) = 0$

$u = 9 \text{ or } u = 2$

$(x - 2)^{1/4} = 9 \text{ or } (x - 2)^{1/4} = 2$

$x - 2 = 6561 \text{ or } x - 2 = 16$

the solutions are 6563 and 18

78. time after 11 am: x

$[100(x + 1)]^2 + (200x)^2 = 700^2$

$100^2(x^2 + 2x + 1) + 200^2 x^2 = 700^2$

$50{,}000x^2 + 20{,}000x + 10{,}000 = 490{,}000$

$5x^2 + 2x - 48 = 0$

$x = \dfrac{-2 + \sqrt{4 + 960}}{10}$

$x = 2.9 \text{ minutes}$

they are 700 mi apart at about 11:03 am.

80. distance from base of first antenna: x

distance from base of second antenna: 75 - x

$\sqrt{x^2 + 400} + \sqrt{(75 - x)^2 + 625} = 90$

$5625 - 150x + 625 = 8100 - 180\sqrt{x^2 + 400} + 400$

$75 + 5x = 6\sqrt{x^2 + 400}$

$5625 + 750x + 25x^2 = 36x^2 + 14400$

$11x^2 - 750x + 8775 = 0$

$(11x - 585)(x - 15) = 0$

$x = 15 \text{ or } x = \dfrac{1170}{22}$

the solutions are 15 ft or 53.18 ft from the base of the first antenna

82. distance AP: x

$\dfrac{\sqrt{x^2 + 16}}{2} + \dfrac{7 - x}{5} = 3.3$

$5\sqrt{x^2 + 16} + 14 - 2x = 33$

$25x^2 + 400 = 4x^2 + 76x + 361$

$21x^2 - 76x + 39 = 0$

$(21x - 13)(x - 3) = 0$

$x = \dfrac{13}{21}$ or $x = 3$

He walks $7 - x = 6\dfrac{8}{21}$ or 4 miles

84. $x(2 - 3x) + (3x^2 + 4)(2 - 3x) = 0$

$(3x^2 + x + 4)(2 - 3x) = 0$

$x = \dfrac{-1 \pm \sqrt{1 - 48}}{6}$ or $x = \dfrac{2}{3}$

the solutions are $\dfrac{-1 \pm i\sqrt{47}}{6}$ and $\dfrac{2}{3}$

86. $(x - 3)^2 - (x - 3)(x^2 - 5x) = 0$

$(x - 3)(x - 3 - (x^2 - 5x)) = 0$

$-(x - 3)(x^2 - 6x + 3) = 0$

$x = 3$ or $x = \dfrac{6 \pm \sqrt{36 - 12}}{2}$

the solutions are 3 and $3 \pm \sqrt{6}$

88. $2x(3x + 4)^{-3} - 9x^2(3x + 4)^{-4} = 0$

$(3x + 4)^{-4}(2x(3x + 4) - 9x^2) = 0$

$6x^2 + 8x - 9x^2 = 0$

$3x^2 - 8x = 0$

$x = 0$ or $3x - 8 = 0$

the solutions are 0 and $\dfrac{8}{3}$

90. $(6x + 2)^{4/3} - 2x(6x + 2)^{1/3} = 0$

$(6x + 2)^{1/3}(6x + 2 - 2x) = 0$

$(6x + 2)^{1/3}(4x + 2) = 0$

$6x + 2 = 0$ or $4x + 2 = 0$

$x = \dfrac{-1}{3}$ or $x = \dfrac{-1}{2}$

the solutions are $\dfrac{-1}{3}$ and $\dfrac{-1}{2}$

92. $\dfrac{1}{2}x^{-1/2} + \dfrac{3}{2}x^{1/2} = 0$

$\dfrac{1}{2}x^{-1/2}(1 + 3x) = 0$

$1 + 3x = 0$

$x = \dfrac{-1}{3}$

the solution is $\dfrac{-1}{3}$

94. $2(x^3 - 1)^{1/3} + x^3(x^3 - 1)^{-2/3} = 0$

$(x^3 - 1)^{-2/3}[2(x^3 - 1) + x^3] = 0$

$2x^3 - 2 + x^3 = 0$

$3x^3 = 2$

$x = \sqrt[3]{\dfrac{2}{3}}$

the solution is $\sqrt[3]{\dfrac{2}{3}}$

CHAPTER 5

Exercise 5.1

2. a.   about 34,000 feet; at about 33 minutes into the flight

    b.   from about 6 to 113 minutes; from 0 to about 14 minutes into the flight

          and again from about 96 minutes to 120 minutes into the flight.

    c.   about 21,000 feet; about 34,000 feet; at about 9 minutes and again at about

          103 minutes; at about 28 minutes and again at about 58 minutes.

    d.   about 17,000 feet; about 8,000 feet (25,000 to 33,000 feet); about 17,000 feet.

    e.   during the first 10 minutes; during the last 10 minutes.

4. a.   about 40 inches   b.   about 9 years old

    c.   the graph appears to level off at about age 18; clearly, height will not

          continue to increase as age does, since people do not grow indefinitely.

    d.   heredity (tall or short parents), nutrition, age at onset of puberty.

6. a.   probably March and April, since they show a gradual increase

    b.   probably June, July, and August, with a gradual decrease

    c.   September, with an instantaneous increase

8. a. $y = 6 - 2(0)$ b. $0 = 6 - 2x$ c. $y = 6 - 2(-1)$

          $y = 6$ $2x = 6$ $y = 6 + 2$

          $(0,6)$ $x = 3$ $y = 8$

                                     $(3,0)$ $(-1,8)$

10. a. $0 + 2y = 5$ b. $5 + 2y = 5$ c. $-3 + 2y = 5$

          $y = \dfrac{5}{2}$ $2y = 0$ $2y = 8$

          $\left(0, \dfrac{5}{2}\right)$ $y = 0$ $y = 4$

                                     $(5,0)$ $(-3,4)$

12. For $x = -2$, $y = 2(-2) + 6 = 2$; for $x = 0$, $y = 2(0) + 6 = 6$; for $x = 2$, $y = 2(2) + 6 = 10$.

     The desired solutions are $(-2,2)$, $(0,6)$, and $(2,10)$.

14. For $x = 0$, $y = \dfrac{4(0)}{0^2 - 1} = 0$; for $x = 2$, $y = \dfrac{4(2)}{2^2 - 1} = \dfrac{8}{3}$; for $x = 4$, $y = \dfrac{4(4)}{4^2 - 1} = \dfrac{16}{15}$.

The desired solutions are $(0,0)$, $\left(2, \dfrac{8}{3}\right)$, and $\left(4, \dfrac{16}{15}\right)$.

16. For $x = 0$, $y = \dfrac{1}{2}\sqrt{4 - 0^2} = 1$; for $x = 1$, $y = \dfrac{1}{2}\sqrt{4 - 1^2} = \dfrac{\sqrt{3}}{2}$; for $x = 2$, $y = \dfrac{1}{2}\sqrt{4 - 2^2} = 0$.

The desired solutions are $(0,1)$, $\left(1, \dfrac{\sqrt{3}}{2}\right)$, and $(2,0)$.

18. $4x - y = 2$

$4x - 2 = y$

$y = 4x - 2$

when $x = -2$, $y = -10$

when $x = -4$, $y = -18$

20. $3x - xy = 6$

$3x - 6 = xy$

$y = \dfrac{3x - 6}{x}$

when $x = 1$, $y = -3$

when $x = 3$, $y = 1$

22. $x^2y - xy = -5$

$y(x^2 - x) = -5$

$y = \dfrac{-5}{x^2 - x}$

when $x = 2$, $y = \dfrac{-5}{2}$

when $x = 4$, $y = \dfrac{-5}{12}$

24. $x^2y - xy + 3 = 5y$

$3 = 5y - x^2y + xy$

$3 = y(5 + x - x^2)$

$y = \dfrac{3}{5 + x - x^2}$

when $x = -2$, $y = -3$

when $x = 2$, $y = 1$

26. $3 = \dfrac{x}{y^2 + 1}$

$3y^2 + 3 = x$

$y^2 = \dfrac{x - 3}{3}$

$y = \pm\sqrt{\dfrac{x - 3}{3}}$

when $x = -2$, y is undefined

when $x = 1$, y is undefined

28. $5x^2 - 4y^2 = 2$

$5x^2 - 2 = 4y^2$

$y^2 = \dfrac{5x^2 - 2}{4}$

$y = \pm\sqrt{\dfrac{5x^2 - 2}{4}} = \pm\dfrac{\sqrt{5x^2 - 2}}{2}$

when $x = 2$, $y = \pm\dfrac{3\sqrt{2}}{2}$

when $x = 4$, $y = \pm\dfrac{\sqrt{78}}{2}$

30.

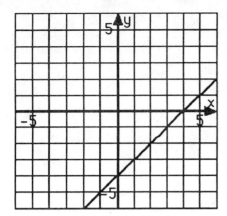

32.

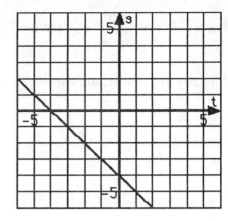

34.

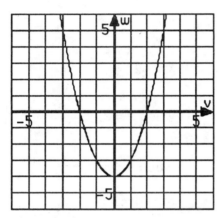

36.

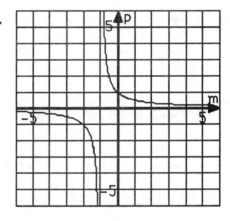

38.

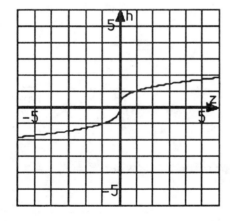

40.

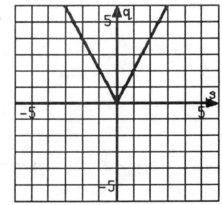

42.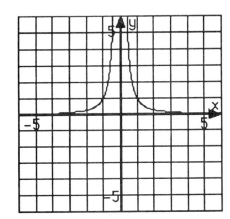

44. a.   I, since income tends to increase with time worked

     b.   I, since temperature rises with the onset of fever and then returns to normal

46.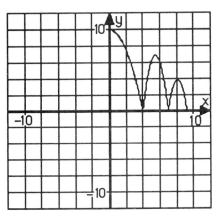

48.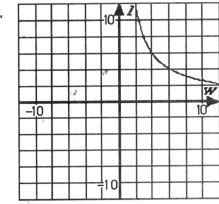

50. a. $y = 3x + 1$ b. $x^3 + 3y^3 = 4$

52. Substitute the values into the equation; if they satisfy the equation, then the point lies on the graph; otherwise it does not.

Exercise 5.2

2. a. $y = 65 + 5t$

    b.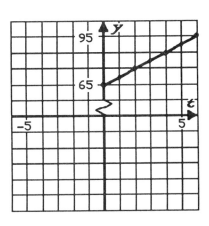

    c.   Noon is t = 6, so $y = 65 + 5(6) = 95°$

    d. $110 = 65 + 5t$

4. a. $g = 20 - \frac{1}{12} m$

    b.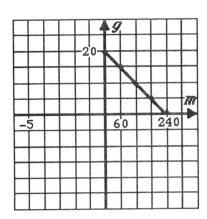

    c. $g = 20 - \frac{1}{12}(100) = 11\frac{2}{3}$ gallons

    d. $\frac{1}{4}$ tank is 5 gallons

71

$$45 = 5t$$

$$t = 9 \text{ hours}$$

9 hours after 6 a.m. is 3 p.m.

$$5 = 20 - \frac{1}{12} m$$

$$\frac{1}{12} m = 15$$

$$m = 180 \text{ miles}$$

6.   a. $s = 20(t - 2.5)$

     b.

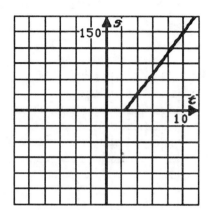

     c. $s = 20(7 - 2.5)$

         $s = 90 \text{ samples}$

     d. $125 = 20(t - 2.5)$

         $6.25 = t - 2.5$

         $t = 8.75 \text{ hours}$

8.   a. $c = 17w - 500$

     b.

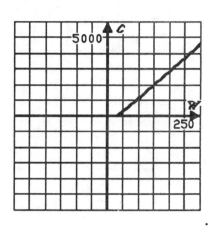

     c. $c = 17(175) - 500$

         $c = 2475 \text{ calories}$

     d. $2200 = 17w - 500$

         $17w = 2700$

         $w \approx 159 \text{ pounds}$

10.   a. $16000 = 600b + 400s$

      b.

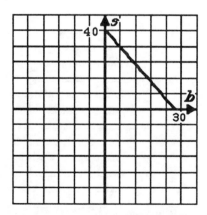

      c. $16000 = 600(10) + 400s$

          $10000 = 400s$

          $s = 25 \text{ hours}$

12.   a. $t = 1706.30 + 0.2a$

      b.

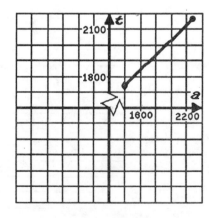

      c. $t = 1706.3 + 0.2(15000 - 13920)$

          $t = 1706.3 + 216 = \$1922.30$

      d. $1800 = 1706.3 + 0.2a$

$0.2a = 93.70$, so $a = 468.5$

income = $468.5 + 13920 =$
$14,388.50

14. If $y = 0$, $2x = 6$ and $x = 3$

If $x = 0$, $-y = 6$ and $y = -6$

The intercepts are $(3,0)$ and $(0,-6)$

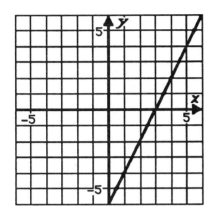

16. If $y = 0$, $2x = 6$ and $x = 3$

If $x = 0$, $6y = 6$ and $y = 1$

The intercepts are $(3,0)$ and $(0,1)$

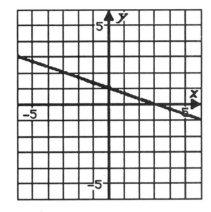

18. If $y = 0$, $\frac{x}{5} = 1$ and $x = 5$

If $x = 0$, $\frac{y}{8} = 1$ and $y = 8$

The intercepts are $(5,0)$ and $(0,8)$

20. If $y = 0$, $\frac{8}{7}x = 1$ and $x = \frac{7}{8}$

If $x = 0$, $-\frac{2}{7}y = 1$ and $y = -\frac{7}{2}$

The intercepts are $\left(\frac{7}{8},0\right)$ and $\left(0,-\frac{7}{2}\right)$

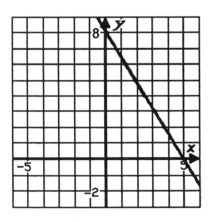

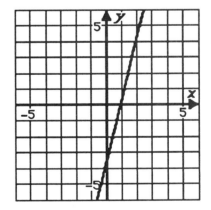

22. If $y = 0$, $30x = 60,000$ and $x = 2000$

If $x = 0$, $45y + 60,000 = 0$ and $y = -1333\frac{1}{3}$

The intercepts are $(2000,0)$ and $(0,-1333\frac{1}{3})$

24. If $y = 0$, $3.2x = 12.8$ and $x = 4$

If $x = 0$, $-0.8y = 12.8$ and $y = -16$

The intercepts are $(4,0)$ and $(0,-16)$

73

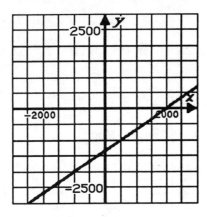
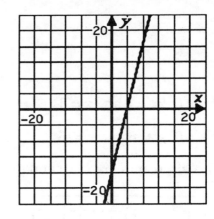

26. The y-intercept = 65° is the temperature at 6 a.m.; the t-intercept = -13 has no physical significance (it would be the time at which the temperature is 0°).

28. The g-intercept = 20 gal is the amount in the tank at the start of the trip; the m-intercept = 240 miles is the distance driven when the tank becomes empty.

30. The s-intercept = -50 has no physical significance; the t-intercept = 2.5 hrs is the time required to prepare the tests.

32. The c-intercept = -500 has no physical significance; the w-intercept = 29.4 lbs is the ideal weight of someone who eats no calories and so has no significance.

34. The b-intercept = $26\frac{2}{3}$ hrs is the amount of biking needed to lose 5 lbs; the s-intercept = 40 hrs is the amount of swimming needed to lose 5 lbs.

36. The t-intercept = $1706.30 is the tax paid by someone who earns $13,920; the a-intercept = -$8,531.50 has no physical significance.

38.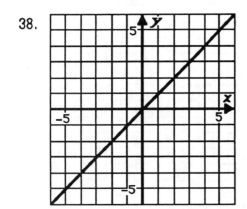

40.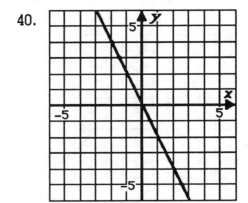

42.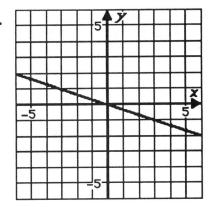

44. a. w = 12h

    b.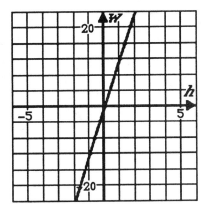

46.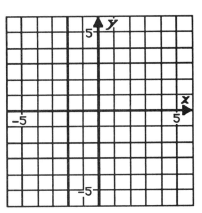

48.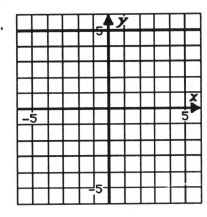

50.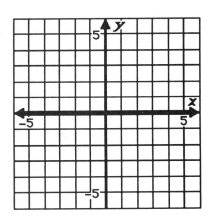

52. $\dfrac{x}{6} + \dfrac{y}{3} = 1$

54. x-intercept = P

    y-intercept = Q

56.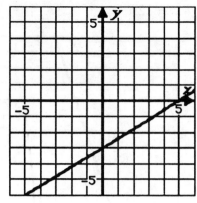

$$\text{area} = \frac{1}{2}\,bh = \frac{1}{2}(5)3 = \frac{15}{2} \text{ square units}$$

58. Find the x-intercept by setting y to 0 and solving for x, and find the y-intercept by setting x to 0 and solving for y. Plot the intercepts and draw a straight line through them.

60. x = k is a vertical line because the equation is satisfied exactly by those points with x-coordinate k, which all lie k units from the y-axis and therefore form a vertical line.

Exercise 5.3

2. a.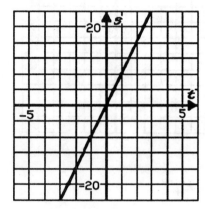

b. $$\frac{\Delta s}{\Delta t} = \frac{\$8 - \$0}{1 \text{ hr} - 0 \text{ hr}} = \frac{\$8}{1 \text{ hr}}$$

c. the slope gives the pay rate of $8.00 per hour.

4. $$m = \frac{9 - (-3)}{1 - (-4)} = \frac{12}{5}$$

6. $$m = \frac{1 - (-4)}{-1 - 5} = \frac{5}{-6} = -\frac{5}{6}$$

8. $$m = \frac{9 - 9}{4 - (-2)} = \frac{0}{6} = 0$$

10. $$m = \frac{-6.2 - (-2.1)}{-3.5 - (-9.7)} = \frac{-4.1}{6.2} = -0.6613$$

12. $$m = \frac{-\frac{4}{3} - \left(-\frac{1}{3}\right)}{\frac{1}{10} - \left(-\frac{7}{5}\right)} = \frac{-1}{\frac{15}{10}} = -\frac{2}{3}$$

14. $$m = \frac{-1\frac{1}{4} - \left(-2\frac{3}{4}\right)}{-\frac{3}{4} - (-3)} = \frac{\frac{3}{2}}{\frac{9}{4}} = \frac{2}{3}$$

16. $$m = \frac{2.7 - 6.9}{2.1 - (-1.9)} = \frac{-4.2}{4} = -1.05$$

18. $$m = \frac{500 - 500}{200 - (-400)} = \frac{0}{600} = 0$$

20. a.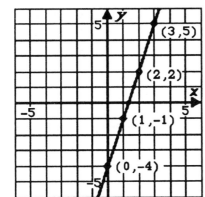

22. a.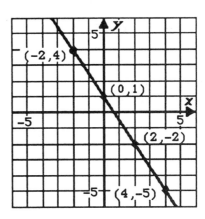

b. $m = \dfrac{2 - (-4)}{2 - 0} = \dfrac{6}{2} = 3$

$m = \dfrac{5 - (-1)}{3 - 1} = \dfrac{6}{2} = 3$

b. $m = \dfrac{1 - (-2)}{0 - 2} = \dfrac{3}{-2} = -\dfrac{3}{2}$

$m = \dfrac{-5 - 4}{4 - (-2)} = \dfrac{-9}{6} = -\dfrac{3}{2}$

24. a. Not linear   b. Not linear   c. Linear

$m_1 = \dfrac{\frac{5}{2} - 5}{2 - 1} = -\dfrac{5}{2}$ $m_1 = \dfrac{9.7 - 6.2}{20 - 10} = 0.35$ m = -5 for all pairs of points

$m_2 = \dfrac{\frac{5}{3} - \frac{5}{2}}{3 - 2} = -\dfrac{5}{6}$ $m_2 = \dfrac{12.6 - 9.7}{30 - 20} = 0.29$

d. Linear $m = \dfrac{5}{2}$ for all pairs of points

26. a. l_1 has positive slope

l_2 has positive slope

l_3 has undefined slope

l_4 has negative slope

b. The lines in order of increasing slope are l_4, l_2, l_1, l_3 (assuming undefined slope is infinite)

28. $y - (-1) = 4(x - (-6))$

$y + 1 = 4x + 24$

$4x - y = -23$

30. $y - 2 = -\dfrac{3}{2}(x - (-1))$

$2y - 4 = -3x - 3$

$3x + 2y = 1$

32. $y - (-1.3) = 1.55(x - 7.2)$

$y + 1.3 = 1.55x + 11.16$

$1.55x - y = -9.86$

34. $m = \dfrac{-3 - (-1)}{2 - 5} = \dfrac{-2}{-3} = \dfrac{2}{3}$

$y - (-1) = \dfrac{2}{3}(x - 5)$

36. $m = \dfrac{3 - 3}{4 - (-1)} = 0$

$y - 3 = 0(x - (-1))$

38. $m = \dfrac{-6.8 - (-18.3)}{5.1 - 11.2}$

$m = -1.885$

77

$$3y + 3 = 2x - 10 \qquad\qquad y - 3 = 0 \qquad\qquad y - (-18.3) = -1.885(x - 11.2)$$

$$2x - 3y = 13 \qquad\qquad y = 3 \qquad\qquad 1.885x + y = 2.812$$

40. a. $d = 26g$

42. a. $F = \dfrac{9}{5}C + 32$

b.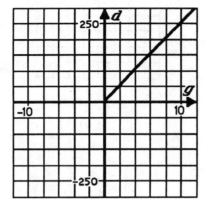

b.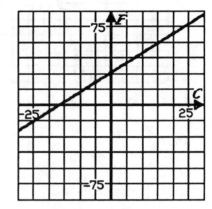

c. $m = \dfrac{26 \text{ miles}}{1 \text{ gal}}$ represents her mileage, 26 mpg.

c. $m = \dfrac{9^\circ F.}{5^\circ C.}$ represents the fact that each degree Celsius = 1.8°F.

44. a. $C = 65t + 125$

46. $m = \dfrac{-6 - 10}{x - 12} = \dfrac{8}{5}$

$-16(5) = 8(x - 12)$

$-80 = 8x - 96$

$8x = 16$

$x = 2$

b.

48. $m = \dfrac{y - (-6)}{2 - 7} = -3$

$-3(-5) = y + 6$

$y + 6 = 15$

$y = 9$

c. $m = \dfrac{\$65}{1 \text{ hr}}$ is the hourly rate of \$65 per hour.

50. a. $\dfrac{2}{\Delta x} = -\dfrac{4}{5}$

b. $\dfrac{-12}{\Delta x} = -\dfrac{4}{5}$

c. $\dfrac{5}{\Delta x} = -\dfrac{4}{5}$

$10 = -4\Delta x$

$-60 = -4\Delta x$

$25 = -4 \Delta x$

horizontal change is $\dfrac{-5}{2}$

horizontal change is 15

horizontal change is $-6\dfrac{1}{4}$

52. 12 miles = 12(5280) feet = 63,360 ft

$$\frac{\Delta y}{63360} = \frac{1}{25}$$

$$25\Delta y = 63360$$

$$\Delta y = 2534.4$$

the elevation changes by 2,534.4 ft.

54. Linear equations have graphs that are straight lines.

56. Since all points on a horizontal line have the same y-coordinate, $\Delta y = 0$ and the slope is 0.
Since all points on a vertical line have the same x-coordinate, $\Delta x = 0$ and the slope is undefined.

Exercise 5.4

2. $3x - y = 7$

$$3x - 7 = y$$

$$y = 3x - 7$$

slope is 3

y-intercept is -7

4. $5x - 4y = 0$

$$5x = 4y$$

$$y = \frac{5}{4}x + 0$$

slope is $\frac{5}{4}$

y-intercept is 0

6. $\frac{7}{6}x - \frac{2}{9}y = 3$

$$\frac{7}{6}x - 3 = \frac{2}{9}y$$

$$y = \frac{21}{4}x - \frac{27}{2}$$

slope is $\frac{21}{4}$

y-intercept is $-\frac{27}{2}$

8. $0.8x + 0.004y = 0.24$

$$-0.8x + 0.24 = 0.004y$$

$$y = -200x + 60$$

slope is -200

y-intercept is 60

10. $y - 37 = 0$

$$y = 37$$

$$y = 0x + 37$$

slope is 0

y-intercept is 37

12. $80x - 360y = 6120$

$$80x - 6120 = 360y$$

$$y = \frac{2}{9}x - 17$$

slope is $\frac{2}{9}$

y-intercept is -17

14. a.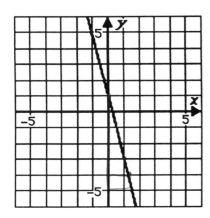

b. $y = -4x + 1$

16. a.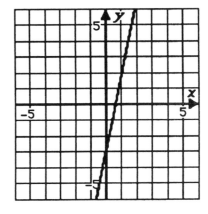

b. $y = 5x - 3$

18. a.

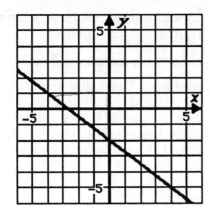

20. a.

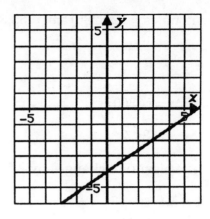

    b. $y = -\dfrac{3}{4}x - 2$

                                                            b. $y = \dfrac{2}{3}x - 4$

22. a. $m \approx \dfrac{9000}{40} = \dfrac{900}{4},\ b \approx -400$ b. $y = \dfrac{900}{4}x - 400$

24. a. $m \approx \dfrac{-4000}{6} = \dfrac{-2000}{3},\ b \approx 10{,}000$ b. $S = \dfrac{-2000}{3}t + 10{,}000$

26. a. $m \approx \dfrac{-0.08}{0.1} = -0.8,\ b \approx 0$ b. $T = -0.8h$

28. a. $2x - 7y = 14$; $7x - 2y = 14$ b. $x + y = 6$; $x - y = 6$

           $2x - 14 = 7y$; $7x - 14 = 2y$ $y = -x + 6$; $x - 6 = y$

           $y = \dfrac{2}{7}x - 2$; $y = \dfrac{7}{2}x - 7$ $-1 = \dfrac{-1}{1}$ perpendicular

           $\dfrac{2}{7} \neq \dfrac{7}{2}$; $\dfrac{2}{7} \neq \dfrac{-1}{\frac{7}{2}}$;   neither   d. $\dfrac{1}{4}x - \dfrac{3}{4}y = \dfrac{2}{3}$; $\dfrac{1}{6}x = \dfrac{1}{2}y + \dfrac{1}{3}$

                                                                             $\dfrac{1}{4}x - \dfrac{2}{3} = \dfrac{3}{4}y$; $\dfrac{1}{6}x - \dfrac{1}{3} = \dfrac{1}{2}y$

    c. $x = -3$; $3y = 5$ $y = \dfrac{1}{3}x - \dfrac{8}{9}$; $y = \dfrac{1}{3}x - \dfrac{2}{3}$

           vertical;   horizontal $\dfrac{1}{3} = \dfrac{1}{3}$;   parallel

           perpendicular

30.   The slope of $\overline{PQ} = \dfrac{8 - 3}{-3 - (-1)} = \dfrac{5}{-2} = -\dfrac{5}{2}$; the slope of $\overline{PR} = \dfrac{5 - 3}{4 - (-1)} = \dfrac{2}{5}$. Since

    $-\dfrac{5}{2} = \dfrac{-1}{\frac{2}{5}}$, $\overline{PQ}$ is perpendicular to $\overline{PR}$ and $\triangle PQR$ is a right triangle.

32. The slope of $\overline{PQ} = \dfrac{4 - (-11)}{-5 - 7} = \dfrac{15}{-12} = -\dfrac{5}{4}$; that of $\overline{QR} = \dfrac{25 - (-11)}{12 - 7} = \dfrac{36}{5}$; that of $\overline{RS}$ $= \dfrac{40 - 25}{0 - 12} = -\dfrac{5}{4}$; that of $\overline{SP} = \dfrac{40 - 4}{0 - (-5)} = \dfrac{36}{5}$. Since the slopes of $\overline{PQ}$ and $\overline{RS}$ are equal, and the slopes of $\overline{QR}$ and $\overline{SP}$ are equal, then $\overline{PQ}$ and $\overline{RS}$ are parallel, and $\overline{QR}$ and $\overline{SP}$ are parallel, and P, Q, R, and S form a parallelogram.

34. $2y - 3x = 5$

$2y = 3x + 5$

$y = \dfrac{3}{2}x + \dfrac{5}{2}$

the slope is $\dfrac{3}{2}$

a parallel line has slope $\dfrac{3}{2}$

$y - 2 = \dfrac{3}{2}(x - (-3))$

$2y - 4 = 3x + 9$

$3x - 2y = -13$

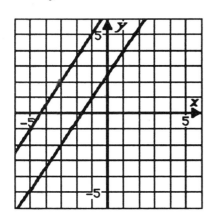

36. $x - 2y = 5$

$x - 5 = 2y$

$y = \dfrac{1}{2}x - \dfrac{5}{2}$

the slope is $\dfrac{1}{2}$

a perpendicular line has slope -2

$y - (-3) = -2(x - 4)$

$y + 3 = -2x + 8$

$2x + y = 5$

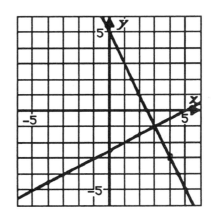

38. Slope of $\overline{PQ} = \dfrac{7 - (-1)}{2 - 4} = \dfrac{8}{-2} = -4$

Slope of $\overline{ST} = \dfrac{k - 4}{5 - (-3)} = \dfrac{k - 4}{8}$

$\dfrac{k - 4}{8} = \dfrac{-1}{-4} = \dfrac{1}{4}$

$4(k - 4) = 8(1) = 8$

$k - 4 = 2$

$k = 6$

40.
a. The acute angles of a right triangle are complementary

b. Same reason as a.

c. Complements of the same angle are equal

d. Same reason as a.

e. Same reason as c.

f. Two triangles are similar if 2 angles of each are equal

g. Definition of slope

81

42.　$ax + by = c;\quad px + qy = r$

　$ax - c = -by;\quad px - r = -qy$

　$y = \frac{-a}{b}x + \frac{c}{b};\quad y = \frac{-p}{q}x + \frac{r}{q}$

Since the lines are perpendicular, $\dfrac{-a}{b} = \dfrac{-1}{\frac{-p}{q}} = \dfrac{q}{p}$, so $-ap = bq$ and $ap + bq = 0$.

44.　Point-slope form is $y - y_1 = m(x - x_1)$ and is used to find the equation of a line when the slope and a point on the line are known. Slope-intercept form is $y = mx + b$ and is used either to find the equation of a line when its slope and y-intercept are known, or (more commonly) to find its slope and y-intercept when when its equation is known.

$ax + by = c$　　$px + qy = r$

$ax - c = -by$

CHAPTER 6

Exercise 6.1

2. Distance traveled is a function of time traveled at a constant speed, since a given time determines a unique distance.

4. The price of a ticket is not a function of the distance traveled, since two flights of the same length could have different prices.

6. Assuming the paycheck involved is subject to Social Security taxes, the amount withheld is a function of the amount of the paycheck, since the amount withheld is set by law by a fixed formula based on earnings.

8. Independent variable: stock number of an item; dependent variable: the number of that item in stock. This determines a function, since each item has a unique quantity available in stock.

10. There are several possible answers here:
(1) Independent variable: persons named in the will; dependent variable: the value of items willed to each person. This is a function, since the items willed to a person have a unique value.
(2) Independent variable: items mentioned in the will; dependent variable: person an item is willed to. This is a function, since each item is willed to one person.
(3) Independent variable: people named in the will; dependent variable: items mentioned in the will. This is not a function, since one person could be willed more than one item.

12. Independent variable: names of people; dependent variable: addresses (or phone numbers). This is not a function, since a person can have more than one address or telephone number (home and business, for example).

14. Independent variable: numbers on the dial; dependent variable: names of radio stations. This is a function, since each number on the dial tunes in one station (assuming distant stations cannot be received).

16. w is a function of y, since each w-value has a unique corresponding y-value.

18. t is NOT a function of s, since t = 10 and t = 30 both correspond to s = 4.

20. d is a function of p, since each p-value has a unique corresponding d-value.

22. Wavelength is a function of frequency, since each value of f has a unique corresponding value of w.

24. The given data determine U as a function of I, since each value of I has a unique corresponding value of U; however, unemployment is not a function of inflation in general, since two years with the same inflation rate could have different unemployment rates.

26. Shipping charge is a function of the cost of merchandise, since the value of m uniquely determines C.

28. a. The independent variable is x, the dependent variable is S.

  b.

x	S
0	20,000
100	20,001
1,000	20,010
5,000	20,050
100,000	21,000

  c. S is a function of x

30. a. The independent variable is x, the dependent variable is P.

  b.

x	P
10,000	-6,500,000
20,000	-8,500,000
50,000	9,500,000
60,000	23,500,000
80,000	63,500,000

  c. P is a function of x

32. a. The independent variable is t, the dependent variable is M.

  b.

t	M
1	600,000
5	120,000
10	60,000
20	30,000
30	20,000

  c. M is a function of t

34. a. The independent variable is h, the dependent variable is d.

  b.

h	d
1	1.22
4	2.44
16	4.88
25	6.1
100	12.2

  c. d is a function of h

36. a. $h(20) = 15.0$ b. $h(60) = 5.0$ c. $x = 30$

38. a. $c(11.50) = 3.75$ b. $c(47.24) = 5.95$ c. $75.01 \leq x \leq 100.00$

40. a. $g(1) = 5(1) - 3 = 2$ b. $g(-4) = 5(-4) - 3 = -23$

  c. $g(14.1) = 5(14.1) - 3 = 67.5$ d. $g\left(\frac{3}{4}\right) = 5\left(\frac{3}{4}\right) - 3 = \frac{3}{4}$

42. a. $r(2) = 2(2) - 2^2 = 0$ b. $r(-4) = 2(-4) - (-4)^2 = -24$

c. $r\left(\dfrac{1}{3}\right) = 2\left(\dfrac{1}{3}\right) - \left(\dfrac{1}{3}\right)^2 = \dfrac{5}{9}$ d. $r(-1.3) = 2(-1.3) - (-1.3)^2 = -4.29$

44. a. $F(0) = \dfrac{1 - 0}{2(0) - 3} = -\dfrac{1}{3}$ b. $F(-3) = \dfrac{1 - (-3)}{2(-3) - 3} = -\dfrac{4}{9}$

c. $F\left(\dfrac{5}{2}\right) = \dfrac{1 - \dfrac{5}{2}}{2\left(\dfrac{5}{2}\right) - 3} = \dfrac{-\dfrac{3}{2}}{2} = -\dfrac{3}{4}$ d. $F(9.8) = \dfrac{1 - 9.8}{2(9.8) - 3} = \dfrac{-8.8}{16.6} = -0.5301$

46. a. $D(4) = \sqrt{5 - 4} = 1$ b. $D(-3) = \sqrt{5 - (-3)} = \sqrt{8} = 2\sqrt{2}$

c. $D(-9) = \sqrt{5 - (-9)} = \sqrt{14}$ d. $D(4.6) = \sqrt{5 - 4.6} = \sqrt{0.4}$

48. a. $T(27) = -3(27)^{2/3} = -27$ b. $T\left(\dfrac{1}{8}\right) = -3\left(\dfrac{1}{8}\right)^{2/3} = -\dfrac{3}{4}$

c. $T(20) = -3(20)^{2/3} = -3\sqrt[3]{400} = -6\sqrt[3]{50}$ d. $T(1000) = -3(1000)^{2/3} = -300$

50. a. $h(2a) = 2(2a)^2 + 6(2a) - 3 = 8a^2 + 12a - 3$

b. $h(a + 3) = 2(a + 3)^2 + 6(a + 3) - 3 = 2(a^2 + 6a + 9) + 6a + 18 - 3 =$

$2a^2 + 12a + 18 + 6a + 18 - 3 = 2a^2 + 18a + 33$

c. $h(a) + 3 = 2a^2 + 6a - 3 + 3 = 2a^2 + 6a$

d. $h(-a) = 2(-a)^2 + 6(-a) - 3 = 2a^2 - 6a - 3$

52. a. $f(4) = -3$ b. $f(-3) = -3$ c. $f(b - 2) = -3$ d. $f(-t) = -3$

54. a. $Q(2t) = 5(2t)^3 = 5(8t^3) = 40t^3$ b. $2Q(t) = 2(5t^3) = 10t^3$

c. $Q(t^2) = 5(t^2)^3 = 5t^6$ d. $[Q(t)]^2 = [5t^3]^2 = 25t^6$

56. $S(850,000) = 20,000 + 0.01(850,000) = \$28,500$

58. $P(20,000) = 0.02(20,000)^2 - 800(20,000) - 500,000 = -8,500,000$, a loss of \$8,500,000

60. $M(12) = \dfrac{600,000}{12} = \$50,000$ 62. $d(20,320) = 1.22\sqrt{20,320} = 173.9$ miles

64. a. $f(2) + f(3) = [1 - 4(2)] + [1 - 4(3)] = (-7) + (-11) = -18$

b. $f(2 + 3) = f(5) = 1 - 4(5) = -19$

c. $f(a) + f(b) = [1 - 4a] + [1 - 4b] = 2 - 4a - 4b$

d. $f(a + b) = 1 - 4(a + b) = 1 - 4a - 4b \neq f(a) + f(b)$

85

66.   a. $f(2) + f(3) = [2^2 - 1] + [3^2 - 1] = (4 - 1) + (9 - 1) = 11$

       b. $f(2 + 3) = f(5) = 5^2 - 1 = 24$

       c. $f(a) + f(b) = [a^2 - 1] + [b^2 - 1] = a^2 + b^2 - 2$

       d. $f(a + b) = (a + b)^2 - 1 = a^2 + 2ab + b^2 - 1 \neq f(a) + f(b)$

68.   a. $f(2) + f(3) = \sqrt{6 - 2} + \sqrt{6 - 3} = \sqrt{4} + \sqrt{3} = 2 + \sqrt{3}$

       b. $f(2 + 3) = f(5) = \sqrt{6 - 5} = 1$

       c. $f(a) + f(b) = \sqrt{6 - a} + \sqrt{6 - b}$

       d. $f(a + b) = \sqrt{6 - (a + b)} = \sqrt{6 - a - b} \neq f(a) + f(b)$

70.   a. $f(2) + f(3) = \dfrac{3}{2} + \dfrac{3}{3} = \dfrac{3}{2} + 1 = \dfrac{5}{2}$

       b. $f(2 + 3) = f(5) = \dfrac{3}{5}$

       c. $f(a) + f(b) = \dfrac{3}{a} + \dfrac{3}{b} = \dfrac{3b + 3a}{ab}$

       d. $f(a + b) = \dfrac{3}{a + b} \neq f(a) + f(b)$

72.   a. $f(0) = 3(0)^2 - 6(0) = 0;\ g(0) = 8 + 4(0) = 8$

       b. $0 = 3x^2 - 6x$ c. $0 = 8 + 4x$ d. $3x^2 - 6x = 8 + 4x$

             $0 = 3x(x - 2)$ $-4x = 8$ $3x^2 - 10x - 8 = 0$

             $x = 0 \text{ or } 2$ $x = -2$ $(x - 4)(3x + 2) = 0$

                                                              $x = 4 \text{ or } -\dfrac{2}{3}$

74.   a. $f(0)$ is undefined; $g(0) = 2(0) - 7 = -7$

       b. $0 = \sqrt{x - 3}$ c. $0 = 2x - 7$ d. $\sqrt{x - 3} = 2x - 7$

             $0 = x - 3$ $2x = 7$ $x - 3 = 4x^2 - 28x + 49$

             $x = 3$ $x = \dfrac{7}{2}$ $(x - 4)(4x - 13) = 0$

                                                             $x = 4 \text{ or } \dfrac{13}{4}$

76. a. $f(x + h) = 2 - 4(x + h) = 2 - 4x - 4h$

    b. $f(x + h) - f(x) = (2 - 4x - 4h) - (2 - 4x) = -4h$

    c. $\dfrac{f(x + h) - f(x)}{h} = \dfrac{-4h}{h} = -4$

78. a. $f(x + h) = 5(x + h)^2 = 5x^2 + 10xh + 5h^2$

    b. $f(x + h) - f(x) = (5x^2 + 10xh + 5h^2) - (5x^2) = 10xh + 5h^2$

    c. $\dfrac{f(x + h) - f(x)}{h} = \dfrac{10xh + 5h^2}{h} = \dfrac{h(10x + 5h)}{h} = 10x + 5h$

80. a. $f(x + h) = (x + h)^2 + 2(x + h) - 6 = x^2 + 2hx + h^2 + 2x + 2h - 6$

    b. $f(x + h) - f(x) = (x^2 + 2hx + h^2 + 2x + 2h - 6) - (x^2 + 2x - 6) = 2hx + h^2 + 2h$

    c. $\dfrac{f(x + h) - f(x)}{h} = \dfrac{2hx + h^2 + 2h}{h} = \dfrac{h(2x + h + 2)}{h} = 2x + h + 2$

82. a. $f(x + h) = \dfrac{1}{(x + h) - 4}$

    b. $f(x + h) - f(x) = \dfrac{1}{(x + h) - 4} - \dfrac{1}{x - 4} = \dfrac{(x - 4) - (x + h - 4)}{(x + h - 4)(x - 4)} = \dfrac{-h}{(x + h - 4)(x - 4)}$

    c. $\dfrac{f(x + h) - f(x)}{h} = \dfrac{\dfrac{-h}{(x + h - 4)(x - 4)}}{h} = \dfrac{-h}{h(x + h - 4)(x - 4)} = \dfrac{-1}{(x + h - 4)(x - 4)}$

84. $g(x) = 20 - 4x$ 86. $H(t) = 2t^2$

Exercise 6.2

2. a. $G(-4) = 4, G(-1) = 4,$
      $G(4)$ is undefined

    b. $s = 0$ and $s = -3$

    c. x-intercept: 1
      y-intercept: 3

    d. $G(s) = -3$

    e. $s = 2$

4. a. $f(-1) = 3, f(3) = 6$

    b. $t = 0$ and $t = 4$

    c. x-intercept: -2
      y-intercept: 5

    d. maximum value: $f(t) = 6$
      minimum value: $f(t) = -1$

    e. maximum occurs at $t = 3$
      minimum occurs at $t = -4$

6. a. $C(0) = 1, C\left(\dfrac{-\pi}{3}\right) = \dfrac{1}{2}, C(\pi) = -1$

    b. $C\left(\dfrac{\pi}{6}\right) \approx \dfrac{5}{6}$

8. a. $P(-3) = 2, P(-2) = 2, P(1) = -1$

    b. For all values of n in the intervals (-2,-1] and (1,2]

c. $x = \dfrac{-5\pi}{3}$, $x = \dfrac{-\pi}{3}$, $x = \dfrac{\pi}{3}$, and $x = \dfrac{5\pi}{3}$

c. maximum value: $P(n) = 3$
minimum value: $P(n) = -1$

d. maximum value: $C(x) = 1$
minimum value: $C(x) = -1$

d. maximum occurs on the intervals $(-5,-4]$ and $(4,5]$

e. maximum occurs at $x = -2\pi$, $x = 0$, $x = 2\pi$
minimum occurs at $x = \pi$, $x = -\pi$

minimum occurs on the interval $(-1,1]$

10.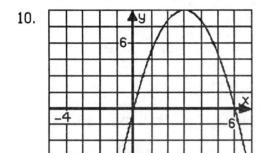

12.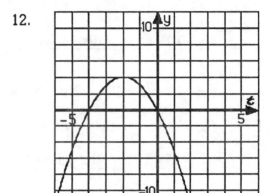

14.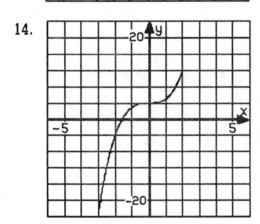

16.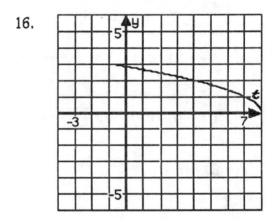

18.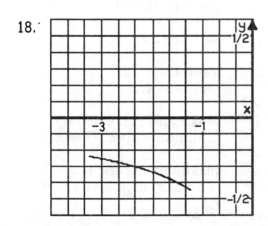

20.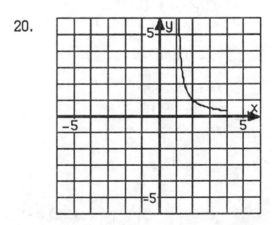

22. domain is $[-5,3]$, range is $[-3,7]$

24. domain is $[-4,5]$, range is $[-1,1) \cup [3,6]$

88

26. domain is $[-2\pi, 2\pi]$, range is $[-1,1]$ 28. domain is $(-5,5]$, range is $\{-1, 0, 2, 3\}$

30. domain is $[-1,7]$, range is $[-7,9]$ 32. domain is $[-6,2]$, range is $[-12,4]$

34. domain is $[-3,2]$, range is $[-23,12]$ 36. domain is $[-1,8]$, range is $[0,3]$

38. domain is $\left[-\dfrac{13}{4}, -\dfrac{5}{4}\right]$, range is $\left[-\dfrac{4}{9}, -\dfrac{4}{17}\right]$

40. domain is $(1,4]$, range is $\left[\dfrac{1}{3}, +\infty\right)$

42. does not represent a function 44. represents a function

46. represents a function 48. does not represent a function

50. represents a function

Exercise 6.3

2.
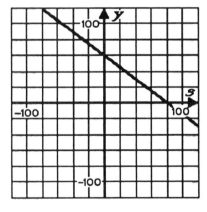

4.
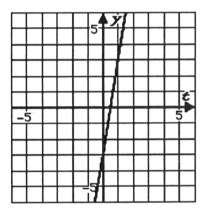

6.
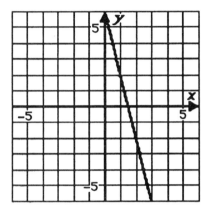

8.
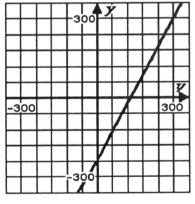

10.

x	y
-3	-27
-2	-8
-1	-1
0	0
1	1
2	8
3	27

12.

x	y
-27	-3
-8	-2
-1	-1
0	0
1	1
8	2
27	3

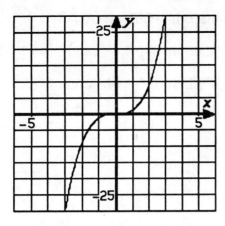

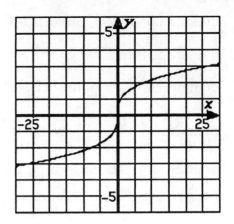

domain: all real numbers
range: all real numbers

domain: all real numbers
range: all real numbers

14.

x	y
-3	$\frac{1}{9}$
-2	$\frac{1}{4}$
-1	1
1	1
2	$\frac{1}{4}$
3	$\frac{1}{9}$

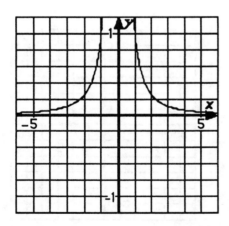

domain: all real numbers except 0
range: all positive real numbers

16.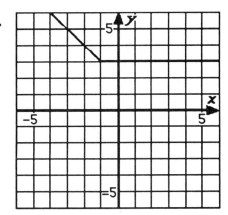

18.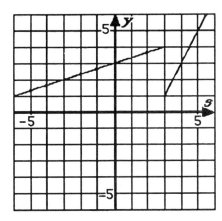

20.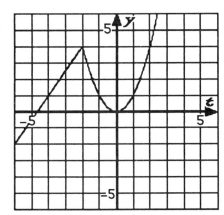

22.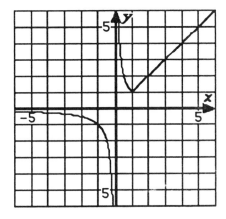

24.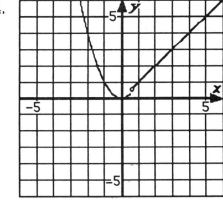

26.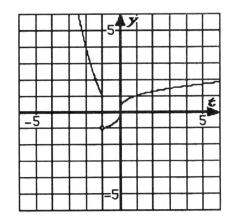

9 1

28. a.

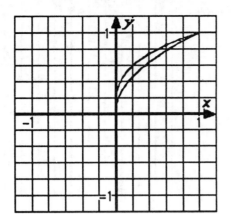

b.

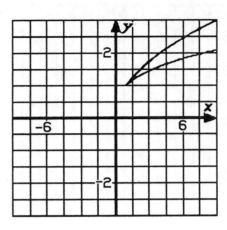

$g(x) = \sqrt[3]{x}$ is greater

$f(x) = \sqrt{x}$ is greater

c.

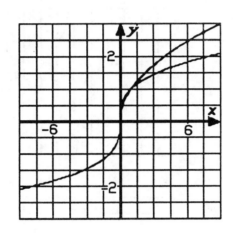

30. a.

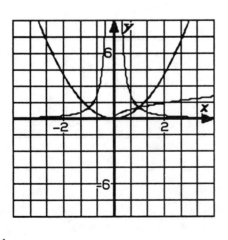

b.

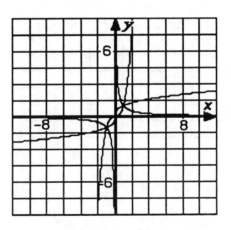

Exercise 6.4

2. a. $C(n) = 12000 + 24n$ b. $C(8000) = 12000 + 24(8000) = \$204,000$

4. a. $C(w) = 5.50(6w) + 2(60)$ b. $C(140) = 5.50(6(140)) + 120 = \$4740$

6.   a.   Since $l = 2\pi r$, we have $r = \dfrac{1}{2\pi}$, and $A(l) = \pi \left(\dfrac{1}{2\pi}\right)^2 = \dfrac{l^2}{4\pi}$

     b. $A(72) = \dfrac{72^2}{4\pi} \approx 412.53$ square inches

8.   a. $r = \dfrac{1}{2}s$ b. $r = \dfrac{1}{2}(30) = 15$ inches

10.   a. $V(d) = \dfrac{1}{3}\pi r^2(3r) = \dfrac{1}{8}\pi d^3$ b. $V(3) = \dfrac{1}{8}\pi(3)^3 = 10.6$ cubic inches

12.   a. $A(r) = 8500\left(1 + \dfrac{r}{100}\right)^5$ b. $A(11) = 8500\left(1 + \dfrac{11}{100}\right)^5 = \$14{,}322.99$

14.   a. $I(x) = 0.09x + 0.15(25{,}000 - x) = 3750 - 0.06x$ b. $I(10{,}000) = \$3150$

16.   a. $A(l) = 2\left(\dfrac{10}{7}\sqrt{53}\right)l$ b. $A(25) = 2\left(\dfrac{10}{7}\sqrt{53}\right)25 = \dfrac{500}{7}\sqrt{53} \approx 520$ sq.ft.

18.   a. $D(t) = \sqrt{(100 - 40t)^2 + (15t)^2} = \sqrt{10{,}000 - 8{,}000t + 1825t^2}$

     b. $D(2) = \sqrt{1300} \approx 36$ feet

20.   a. $y = kx^3$ 22.   a. $y = \dfrac{k}{x^2}$

         $120 = k(2)^3 = 8k$ $3.6 = \dfrac{k}{25}$

         $k = 15$ $k = 25(3.6) = 90$

         $y = 15x^3$ $y = \dfrac{90}{x^2}$

     b. $y = 15(20)^3$ b. $y = \dfrac{90}{(0.4)^2}$

         $y = 120{,}000$ $y = 562.5$

     c.   c.

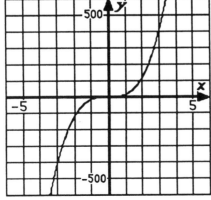
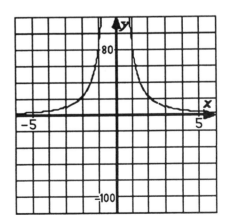

24. a. $L = kT^2$

$\dfrac{13}{4} = k(2)^2 = 4k$

$k = \dfrac{13}{16}$

$L = \dfrac{13}{16} T^2$

b. $L = \dfrac{13}{16}(17)^2$

$L = 234.8125$ feet

c.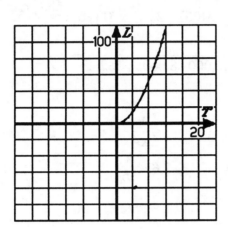

26. a. $F = \dfrac{k}{L}$

$100 = \dfrac{k}{3}$

$k = 300$

$F = \dfrac{300}{L}$

b. $F = \dfrac{300}{4}$

$F = 75$ lbs

c.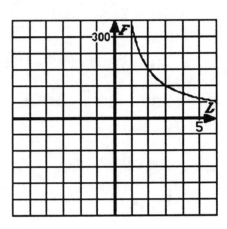

28. a. $V = kD$

$15,000 = k(980,000,000)$

$k = \dfrac{15,000}{980,000,000} = \dfrac{3}{196,000}$

$V = \dfrac{3}{196,000} D$

b. $61,000 = \dfrac{3}{196,000} D$

$D = \dfrac{196,000}{3}(61,000)$

$= 3,985,000,000$ light years

30. a. $I = \dfrac{k}{R}$

$10 = \dfrac{k}{12}$

$k = 12(10) = 120$

$I = \dfrac{120}{R}$

b. $I = \dfrac{120}{9.6}$

$I = 12.5$ amps

94

c.

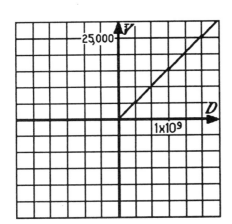

c.

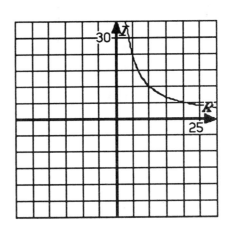

32. $C(w) = 10\left[2w + \dfrac{125}{w}\right]$

34. $A(x) = \left(\dfrac{x}{4}\right)^2 + \dfrac{(100 - x)^2}{4\pi}$

36. $C(x) = 25(50 - 2x)^2 + 10(100x + 2x(50 - 2x)) = 25(50 - 2x)^2 + 10(200x - 4x^2)$

38. $C(x) = 50\sqrt{144 + (18 - x)^2} + 30x$

40. $R = \dfrac{k}{d^2}$

42. $P = kV$

$R' = \dfrac{k}{\left(\dfrac{2}{3}d\right)^2} = \dfrac{k}{\dfrac{4}{9}d^2} = \dfrac{9}{4}\left(\dfrac{k}{d^2}\right) = \dfrac{9}{4}R$

$P' = k(0.7V) = 0.7(kV)$

the new resistance is $\dfrac{9}{4}$ or 2.25 times the old

power decreases by 30%

44. a. $y = \dfrac{k}{x}$, so x varies inversely with y.

    b. $x = ky$, so x varies directly with y.

Exercise 6.5

2.

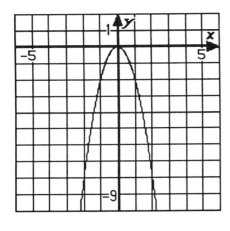

4.

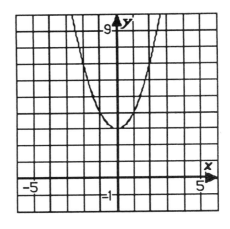

95

6.

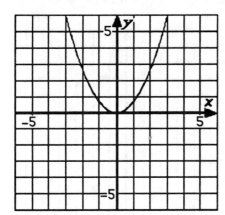

8.

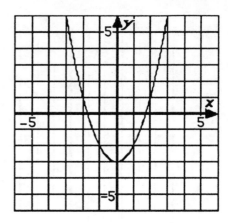

10.

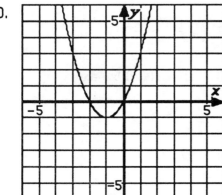

12.

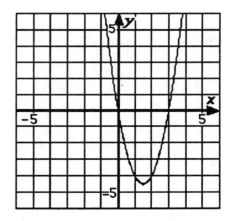

14.

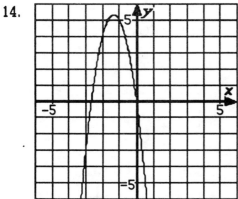

16. $x = \dfrac{-b}{2a} = \dfrac{-5}{-4}$, $y = -2\left(\dfrac{5}{4}\right)^2 + 5\left(\dfrac{5}{4}\right) - 1$

vertex is $\left(\dfrac{5}{4}, \dfrac{17}{8}\right)$

18. $x = \dfrac{-3}{2(-1)} = \dfrac{3}{2}$, $y = 2 + 3\left(\dfrac{3}{2}\right) - \left(\dfrac{3}{2}\right)^2 = \dfrac{17}{4}$

vertex is $\left(\dfrac{3}{2}, \dfrac{17}{4}\right)$

20. $x = \dfrac{-\dfrac{1}{2}}{2\left(\dfrac{-3}{4}\right)} = \dfrac{1}{3}$, $y = \dfrac{-3}{4}\left(\dfrac{1}{3}\right)^2 + \dfrac{1}{2}\left(\dfrac{1}{3}\right) - \dfrac{1}{4} = \dfrac{-1}{6}$, vertex is $\left(-\dfrac{1}{3}, -\dfrac{1}{6}\right)$

22. $x = \dfrac{-(-0.2)}{2(4.6)} = \dfrac{1}{46}$, $y = 5.1 - 0.2\left(\dfrac{1}{46}\right) + 4.6\left(\dfrac{1}{46}\right)^2 = \dfrac{469}{92}$, vertex is $\left(\dfrac{1}{46}, \dfrac{469}{92}\right)$

96

24. Vertex is $\left(-\frac{1}{2}, -\frac{25}{4}\right)$

x-intercepts are -3 and 2

y-intercept is -6

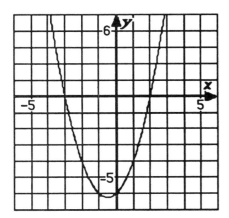

26. Vertex is $\left(\frac{1}{3}, \frac{25}{3}\right)$

x-intercepts are 2 and $-\frac{4}{3}$

y-intercept is 8

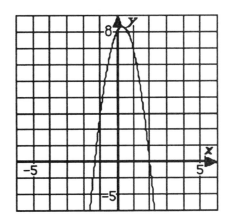

28. Vertex is $\left(\frac{1}{4}, -\frac{25}{32}\right)$

x-intercepts are $\frac{3}{2}$ and -1

y-intercept is $-\frac{3}{4}$

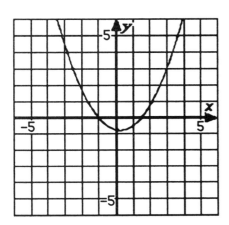

30. Vertex is (3,1)

There are no x-intercepts

y-intercept is 10

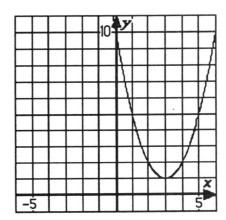

32. Vertex is $\left(\frac{1}{6}, -\frac{23}{12}\right)$

There are no x-intercepts

y-intercept is -2

34. Vertex is (3,-7)

x-intercepts are $3 \pm \sqrt{7}$

y-intercept is 2

97

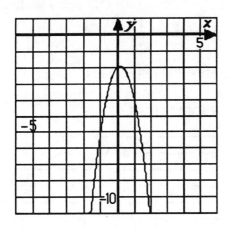

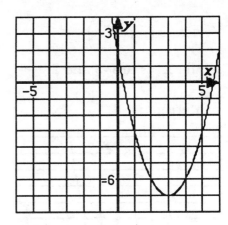

36. Vertex is (-2,3)

x-intercepts are $\dfrac{-4 \pm \sqrt{6}}{2}$

y-intercept is -5

38. **a.** Vertex is (1,4)

The object reaches its maximum height of 4 feet after 1 second.

b.

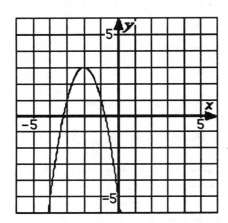

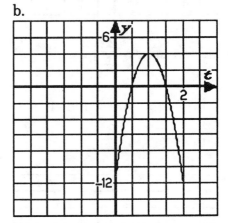

40. **a.** $A(x) = x(40 - x) = -x^2 + 40x$

Vertex is (20,400)

The dimensions are 20 x 20, the area is 400 square yards

42. **a.** Distance to river: x

$A(x) = x(600 - 3x) = 600x - 3x^2$

Vertex is (100, 30,000)

Largest area is 30,000 m^2

b.

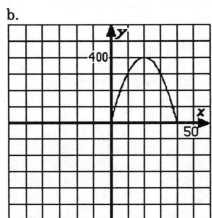

b.

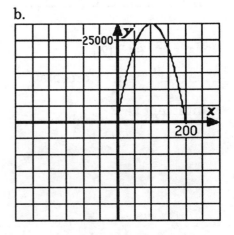

44. a. $R(x) = (2400 - 100x)(16 + x)$

$= -100x^2 + 800x + 38,400$

The vertex is (4, 40,000)

The maximum revenue occurs with $16 + 4 = 20$ people and is $40,000.

b.

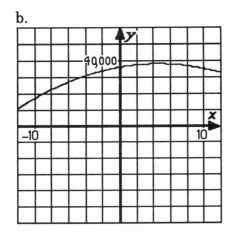

46. a. $R(x) = x(180 - \frac{1}{3}x)$

$= 180x - \frac{1}{3}x^2$

The vertex is (270,24300)

The maximum revenue occurs at a price of $270 per skylight and is $24,300.

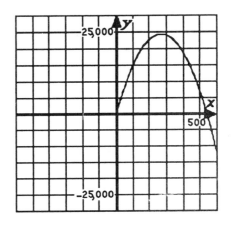

48. $y = x^2 - 3$, $y = 2x^2 - 3$

(all parabolas with equations of the form $y = ax^2 - 3$ are acceptable)

50. $y = x^2 + 2x + 4$, $y = 2x^2 - x + 4$

(all parabolas with equations of the form $y = ax^2 + bx + 4$ are acceptable)

52. $y = x^2 - 3x - 4$, $y = 2x^2 - 6x - 8$

(all parabolas with equations of the form $y = a(x + 1)(x - 4)$ are acceptable)

54. $y = x^2 + 8x + 18$, $y = -x^2 - 8x - 14$

(all parabolas with equations of the form $y = a(x + 4)^2 + 2$ are acceptable)

b. $P(x) = R(x) - C(x)$

$= (180x - \frac{1}{3}x^2) - (60x + 500)$

$P(x) = -500 + 120x - \frac{1}{3}x^2$

Vertex is (180,10300)

Maximum profit occurs at a price of $180 and is $10,300

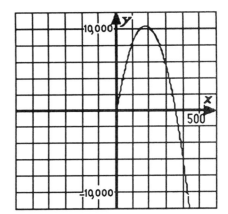

CHAPTER 7

Exercise 7.1

2.

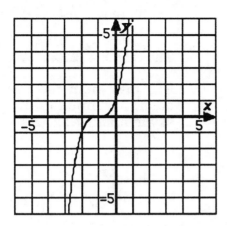

4.

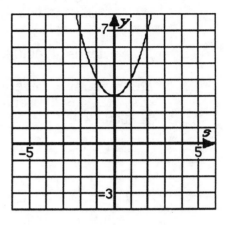

6.

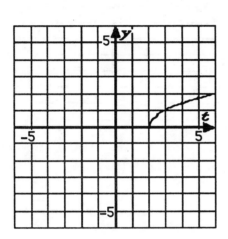

8.

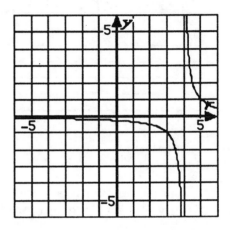

10.

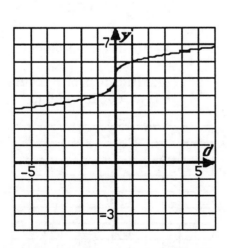

12.

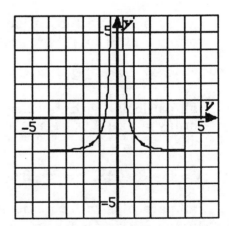

14.

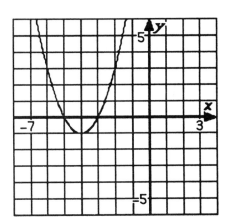

16.

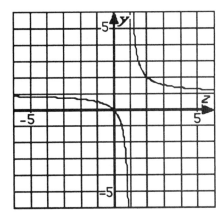

18.

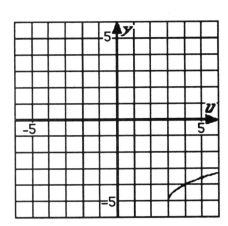

20.

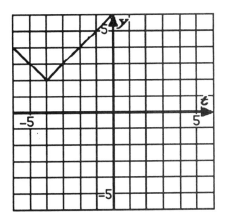

22.

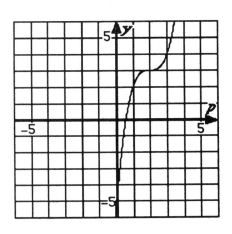

24.

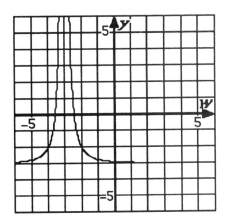

26.

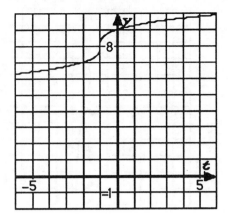

28.

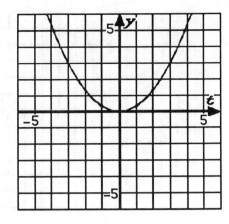

30.

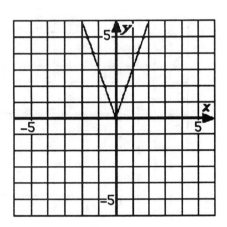

32.

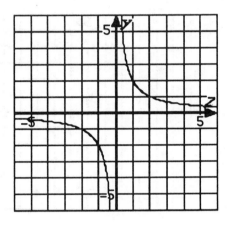

34.

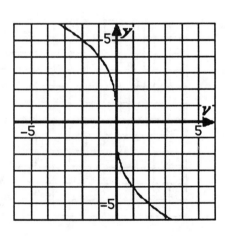

36.

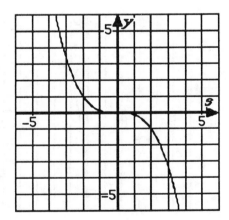

38.

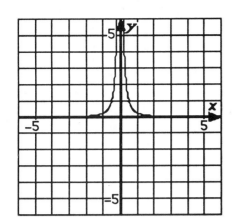

40. $f(x) = \dfrac{-1}{x - 2}$

42. $f(x) = (x + 1)^2 + 1$

44. $f(x) = \dfrac{1}{(x + 3)^2} - 2$

46.   a. $y = (x^2 - 2x + 1) - 2$

               $y = (x - 1)^2 - 2$

   b.

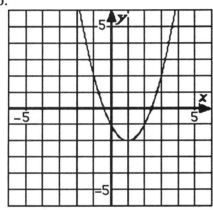

48.   a. $y = (x^2 + 4x + 4) + 1$

               $y = (x + 2)^2 + 1$

   b.

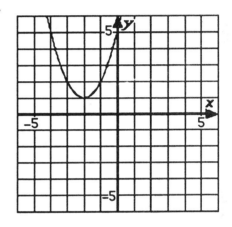

Exercise 7.2

2.

x	y
-3	0
-2	-10
-1	-8
0	0
1	8
2	10
3	0

4.

x	y
-3	0
-2	-15
-1	0
0	9
1	0
2	-15
3	0

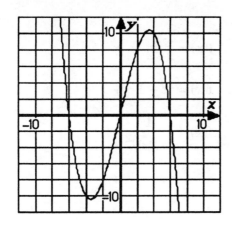
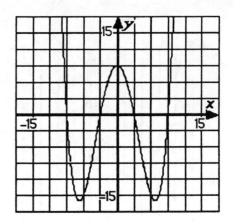

6.
$$\begin{array}{r|rrrrr}
2 & 1 & -10 & 5 & -3 & 6 \\
 & & 2 & -16 & -22 & -50 \\
\hline
 & 1 & -8 & -11 & -25 & -44
\end{array}$$

Q(2) = -44

$$\begin{array}{r|rrrrr}
-3 & 1 & -10 & 5 & -3 & 6 \\
 & & -3 & 39 & -132 & 405 \\
\hline
 & 1 & -13 & 44 & -135 & 411
\end{array}$$

Q(-3) = 411

8.
$$\begin{array}{r|rrrrr}
3 & 4 & -2 & 3 & 0 & -5 \\
 & & 12 & 30 & 99 & 297 \\
\hline
 & 4 & 10 & 33 & 99 & 292
\end{array}$$

T(3) = 292

$$\begin{array}{r|rrrrr}
-4 & 4 & -2 & 3 & 0 & -5 \\
 & & -16 & 72 & -300 & 1200 \\
\hline
 & 4 & -18 & 75 & -300 & 1195
\end{array}$$

T(-4) = 1195

10.

	1	-3	-6	8
-3	1	-6	12	-28
-2	1	-5	4	0
-1	1	-4	-2	10
0	1	-3	-6	8
1	1	-2	-8	0
2	1	-1	-8	-8
3	1	0	-6	-10

12.

	1	5	-1	-5	0
-3	1	2	-7	16	-48
-2	1	3	-7	9	-18
-1	1	4	-5	0	0
0	1	5	-1	-5	0
1	1	6	5	0	0
2	1	7	13	21	42
3	1	8	23	64	192

14.   Q(2.3) = ([-3(2.3) + 3](2.3) + 2)(2.3) - 5 = -21.031

Q(-2.3) = (-3(-2.3) + 3](-2.3) + 2)(-2.3) - 5 = 42.771

16.   Q(3) = [([([0.05(3) + 0](3) - 1.8)(3) + 1](3) + 0)(3) +12](3) + 0 = -46.35

Q(-2) = [([([0.05(-2) + 0](-2) - 1.8)(-2) + 1](-2) + 0)(-2) +12](-2) + 0 = -57.6

18. $x^3 + 3x = x(x^2 + 3)$, and $x^2 + 3 = 0$ means $x^2 = -3$ or $x = \pm i\sqrt{3}$. The zeroes of P(x)

are 0, $i\sqrt{3}$, and $-i\sqrt{3}$.

20. $4x^4 - 9x^2 = x^2(4x^2 - 9) = x^2(2x - 3)(2x + 3)$. By the factor theorem, the zeroes of

P(x) are $0, \frac{3}{2}$, and $\frac{-3}{2}$, with 0 a zero of multiplicity two.

22. $x^3 - 2x^2 + 3x - 6 = x^2(x - 2) + 3(x - 2) = (x^2 + 3)(x - 2)$. Since $x^2 + 3 = 0$ means x =

$\pm i \sqrt{3}$, the zeroes of P(x) are 2, i $\sqrt{3}$, and -i $\sqrt{3}$.

24. $x^4 + 4x^3 + 3x^2 = x^2(x^2 + 4x + 3) = x^2(x + 1)(x + 3)$. By the factor theorem, the

zeroes of P(x) are 0, -1, and -3, with 0 a zero of multiplicity two.

26. Since $P(1) = 2(1)^3 - 5(1)^2 + 4(1) - 1 = 0$, x - 1 is a factor of $2x^3 - 5x^2 + 4x - 1$.

28. Since $P(-1) = 2(-1)^3 - 5(-1)^2 + 4(-1) - 1 = -12$, x + 1 is not a factor of $2x^3 - 5x^2 + 4x - 1$.

30. $(-1)^3 + 2(-1)^2 - 1 = (-1) + 2 - 1 = 0$; since x = -1 is a solution of P(x) = 0, x + 1 is a

factor of P(x). By long division, $P(x) = (x + 1)(x^2 + x - 1)$, and applying the

quadratic formula gives the additional solutions $\frac{-1 \pm \sqrt{5}}{2}$.

32. $(-5)^4 + 5(-5)^3 - (-5)^2 - 5(-5) = 625 - 625 - 25 + 25 = 0$; since x = -5 is a solution of P(x)

= 0, x + 5 is a factor of P(x). By long division, $P(x) = (x + 5)(x^3 - x) =$

$x(x + 5)(x - 1)(x + 1)$, so the remaining solutions are 0, 1, and -1.

Exercise 7.3

2. $g(-2) = -16$
   $g(0) = -32$

4. $F(0) = 9$ $F(2) = 1$
   $F(-1) = 64$ $F(4) = 9$

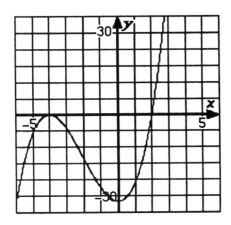

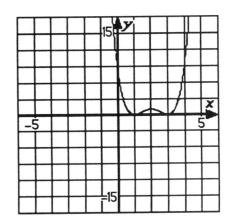

6.   H(-2) = -16   H(0) = 4
      H(3) = 64

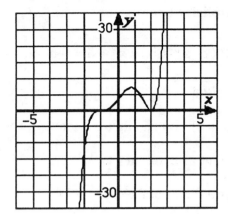

8.   Q(-1) = -24
      Q(2) = -48

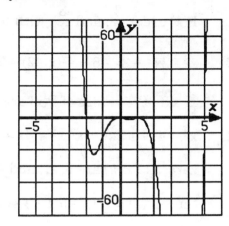

10.   p(-1) = -8   p(0) = -8
      p(2) = 64

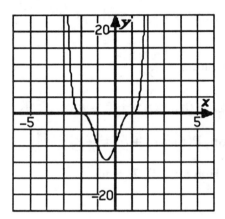

12.   T(x) = x^3 - 4x^2 + 3x = x(x - 3)(x - 1)
      The x-intercepts are 0, 3, and 1

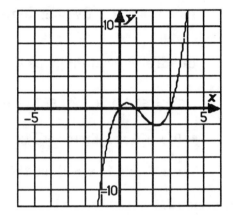

14.   f(x) = 12x - x^3 = x(12 - x^2)

      x = 0 or x^2 = 12

      The x-intercepts are 0, 2 $\sqrt{3}$, and -2 $\sqrt{3}$

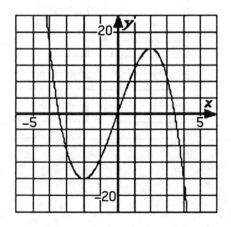

16.   v(x) = x^3 + x^2 - 25x - 25 = x^2(x + 1) -

      25(x + 1) = (x - 5)(x + 5)(x + 1)

      The x-intercepts are 5, -5, and -1

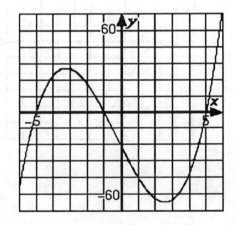

18. $G(x) = x^4 - x^2 = x^2(x - 1)(x + 1)$
The x-intercepts are 0, 1, and -1

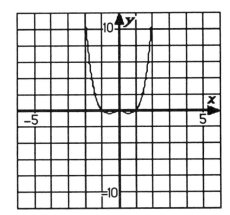

20. $m(x) = x^4 - 15x^2 + 36 =$
$(x^2 - 12)(x^2 - 3)$, so $x^2 = 12$ or $x^2 = 3$.
The x-intercepts are $\pm 2\sqrt{3}$ and $\pm\sqrt{3}$

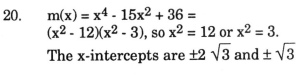
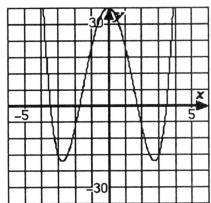

22. $q(x) = x^4 - x^3 + x - 1 =$
$x^3(x - 1) + (x - 1) = (x^3 + 1)(x - 1) =$
$(x + 1)(x - 1)(x^2 - x + 1)$

The zeroes of $q(x)$ are 1, -1, and $\dfrac{1 \pm \sqrt{-3}}{2}$,
so the x-intercepts are 1 and -1

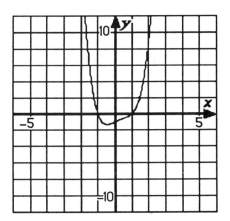

24. $h(x) = x^4 + 3x^3 - x^2 - 3x =$
$x^3(x + 3) - x(x + 3) = (x^3 - x)(x + 3) =$
$x(x - 1)(x + 1)(x + 3)$

The x-intercepts are 0, 1, -1, and -3

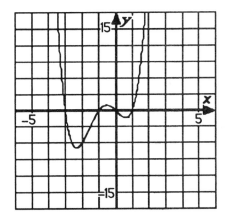

107

26. $s(x) = (x - 3)(x + 3)(x - 1)^2$
       The x-intercepts are 3, -3, and 1

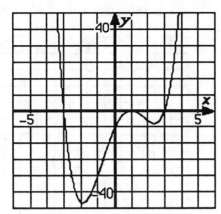

28.

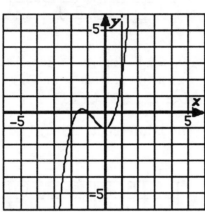

30.

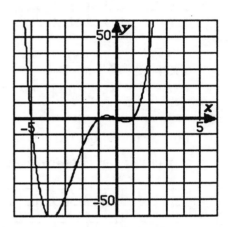

Exercise 7.4

2.   a.   the line x = 4 is a vertical
          asymptote

     b.   the domain is the set of all
          real numbers except 4

6.   a.   the lines x = 1 and x = 2 are
          vertical asymptotes

     b.   the domain is the set of all

          real numbers except 1 and 2

4.   a.   the lines x = -1 and x = 4 are
          vertical asymptotes

     b.   the domain is the set of all
          real numbers except -1 and 4

8.   Since $x^2 + 5x + 4 = (x + 4)(x + 1)$,

     the lines x = -4 and x = -1 are verti-

     cal asymptotes. Since $\dfrac{2x - 4}{x^2+5x+4}$ ÷

     $\dfrac{x^2}{x^2} = \dfrac{\dfrac{2}{x} - \dfrac{4}{x^2}}{1 + \dfrac{5}{x} + \dfrac{4}{x^2}}$ approaches 0, the

     line y = 0 is a horizontal

     asymptote.

10. The line x = 3 is a vertical asymp-

tote. Since $\dfrac{2x + 1}{x - 3} \div \dfrac{x}{x} = \dfrac{2 + \dfrac{1}{x}}{1 - \dfrac{3}{x}}$

approaches 2, the line y = 2 is a

horizontal asymptote.

12. Since $x^2 - x - 12 = (x - 4)(x + 3)$, the

lines x = 4 and x = -3 are vertical

asymptotes. Since $\dfrac{x^2}{x^2 - x - 12} \div \dfrac{x^2}{x^2} =$

$\dfrac{1}{1 - \dfrac{1}{x} - \dfrac{12}{x^2}}$ approaches 1, the line

y = 1 is a horizontal asymptote.

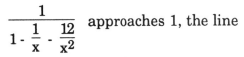

14. There is no x-intercept; the y-

intercept is $\dfrac{-1}{3}$. There is a vertical

asymptote at x = 3 and a horizontal

asymptote at y = 0.

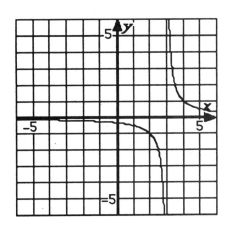

16. There is no x-intercept; the y-

intercept is -2. There are vertical

asymptotes at x = -2 and x = 1 and a

horizontal asymptote at y = 0.

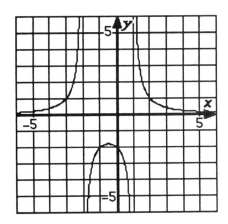

18. There is no x-intercept; the y-intercept is $\frac{-2}{3}$. There are vertical asymptotes at x = 3 and x = -2 and a horizontal asymptote at y =0.

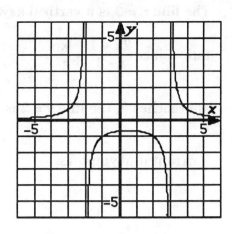

20. The x-intercept and the y-intercept are 0. There is a vertical asymptote at x = 2 and a horizontal asymptote at y = 1.

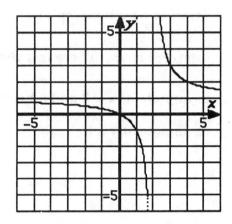

22. The x-intercept is 1, the y-intercept is $\frac{1}{3}$. There is a vertical asymptote at x = 3 and a horizontal asymptote at y = 1.

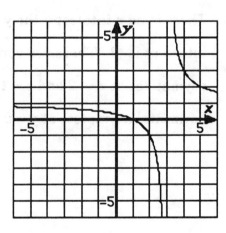

24. The x-intercept and y-intercept are 0. There are vertical asymptotes at x = 3 and x = -3 and a horizontal asymptote at y = 0.

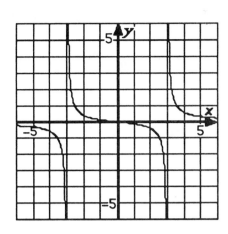

26. The x-intercept is -1; the y-intercept is $\frac{-1}{6}$. There are vertical asymptotes at x = 3 and x = -2 and a horizontal asymptote at y = 0.

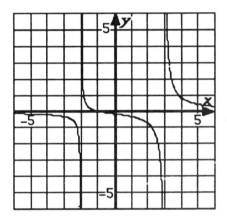

28. The x-intercept and y-intercept are 0. There are vertical asymptotes at x = 1 and x = -1 and a horizontal asymptote at y = 2.

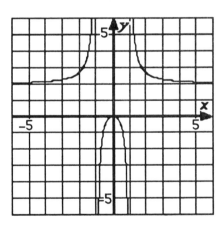

30. There is no x-intercept; the y-intercept is $\frac{1}{2}$. There are no vertical asymptotes and there is a horizontal asymptote at y = 1.

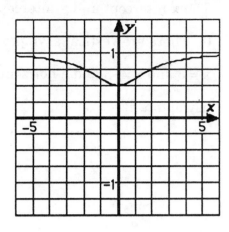

32. a. the domain is the set of all real numbers except -1

    b. $\frac{x^2 - 1}{x + 1} = \frac{(x - 1)(x + 1)}{x + 1} = x - 1$

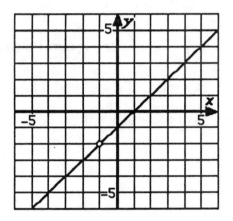

34. a. the domain is the set of all real numbers except 3 and -3

    b. $\frac{x - 3}{x^2 - 9} = \frac{x - 3}{(x - 3)(x + 3)} = \frac{1}{x + 3}$

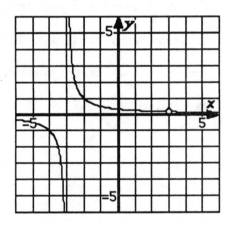

Exercise 7.5

2. $x = y - 3$, so the inverse is $y = x + 3$

4. $x = \frac{y}{5}$, so the inverse is $y = 5x$

6. $x = 3y + 1$, so the inverse is $y = \frac{x - 1}{3}$

8. $x = \frac{5 - y}{3}$, so the inverse is $y = 5 - 3x$

10. $x = y^3 - 8$, so the inverse is $y = \sqrt[3]{x + 8}$

12. $x = \frac{1}{y}$, so the inverse is $y = \frac{1}{x}$

14. $x = \sqrt[3]{y + 1}$, so the inverse is $y = x^3 - 1$

16. $x = \frac{1}{y} - 3$, so the inverse is $y = \frac{1}{x + 3}$

112

18.   a. $x = \dfrac{2}{y+1}$, so $y = \dfrac{2}{x} - 1$ and the inverse is $g(x) = \dfrac{2}{x} - 1$

      b. $f(3) = \dfrac{2}{3+1} = \dfrac{1}{2}$, and $g\left(\dfrac{1}{2}\right) = \dfrac{2}{\frac{1}{2}} - 1 = 4 - 1 = 3$

      c. $g(-1) = \dfrac{2}{-1} - 1 = -3$, and $f(-3) = \dfrac{2}{-3+1} = -1$

20.

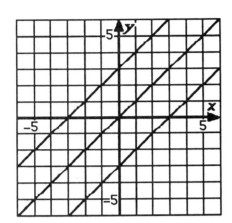

22.

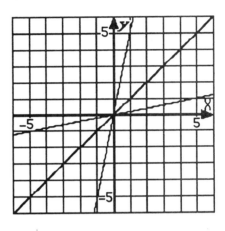

24.

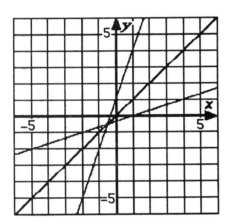

26.

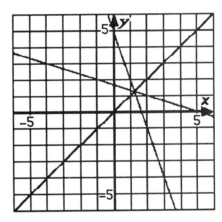

28.

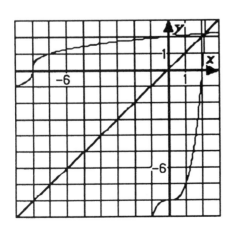

30.

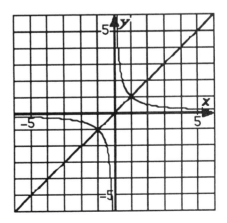

(the function and its inverse have
the same graph)

113

32.

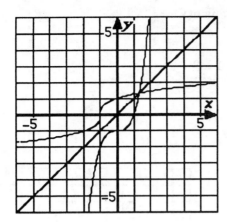

34.

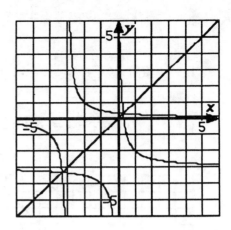

36. Graphs a. and c. fail the horizontal line test and do not have inverses which are also functions; graphs b. and d. pass the horizontal line test and do have inverses which are also functions.

38. a. The function $f(x) = x^3$ has an inverse $g(x) = \sqrt[3]{x}$ which is also a function.

b. The function $f(x) = x^4$ has an inverse $y = \pm \sqrt[4]{x}$ which is not a function.

40. a. The function $f(x) = \sqrt{x}$ has an inverse $g(x) = x^2$ with domain $x \geq 0$ which is a function.

b. The function $f(x) = \sqrt[3]{x}$ has an inverse $g(x) = x^3$ which is also a function.

42. Since $s = \dfrac{y - 3}{4}$, the inverse is $y = 4s + 3$, and $G^{-1}(-2) = 4(-2) + 3 = -5$.

44. Since $z = 1 - 2y^3$, the inverse is $y = \sqrt[3]{\dfrac{1 - z}{2}}$, and $p^{-1}(7) = \sqrt[3]{\dfrac{1 - 7}{2}} = \sqrt[3]{-3}$.

46. $x = \dfrac{3y + 1}{y - 3}$, so $xy - 3x = 3y + 1$, $xy - 3y = 3x + 1$, and $y = f^{-1}(x) = \dfrac{3x + 1}{x - 3}$.

48. a. To find $f^{-1}(7)$, we need to find the x-value for which $f(x) = 7$, that is, for which $7 = x^5 + x^3 + 7$. But $x^5 + x^3 = 0$ means $x^3(x^2 + 1) = 0$, which has only one real solution, 0. Thus $f^{-1}(7) = 0$ (since $f(0) = 7$).

b. As above, we want an x-value for which $f(x) = 5$, that is, $x^5 + x^3 + 7 = 5$, or $x^5 + x^3 + 2 = 0$. Clearly, $(-1)^5 + (-1)^3 + 2 = (-1) + (-1) + 2 = 0$, so -1 is the value we seek. Thus $f^{-1}(5) = -1$ (since $f(-1) = 5$).

114

50. a. f(-1) = y means f⁻¹(y) = -1, so f(-1) = 1.

b. f(1) = y means f⁻¹(y) = 1, so f(1) = -2.

52. a.

x	f(x)
-2	4
-1	2
0	1
1	$\frac{1}{2}$
2	$\frac{1}{4}$

c.

x	f⁻¹(x)
4	-2
2	-1
1	0
$\frac{1}{2}$	1
$\frac{1}{4}$	2

b.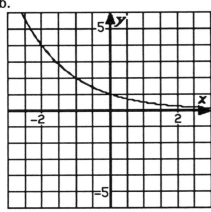

d.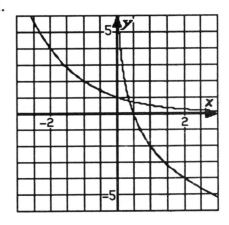

CHAPTER 8

Exercise 8.1

2. a. $P(t) = 24 \cdot 3^t$

    b. $P(6) = 24 \cdot 3^6 = 17{,}496$

    $P\left(\dfrac{3}{4}\right) = 24 \cdot 3^{3/4} = 54$

    c.
    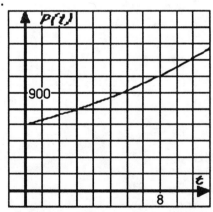

4. a. $P(t) = 800(1.8)^{t/3}$

    b. $P(1) = 800(1.8)^{1/3} = 973$

    $P(10) = 800(1.8)^{10/3} = 5{,}675$

    c.
    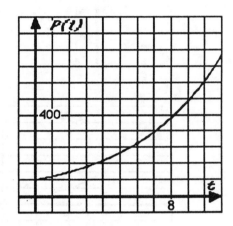

6. a. $P(t) = 600(1.073)^t$

    b. $P(3) = 600(1.073)^3 = 741.23$
    $P(6) = 600(1.073)^6 = 915.69$

    c.
    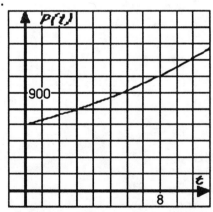

8. a. $P(t) = 135{,}000(1.1)^t$

    b. $P(3) = 135{,}000(1.1)^3 = 179{,}685$
    $P(7) = 135{,}000(1.1)^7 = 263{,}077$

    c.
    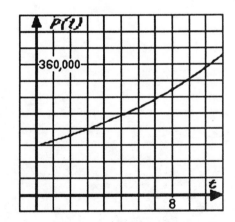

10. a. $P(t) = 500(1.23)^t$

     b. $P(5) = 500(1.23)^5 = 1,407$

           $P(15) = 500(1.23)^{15} = 11,156$

     c.
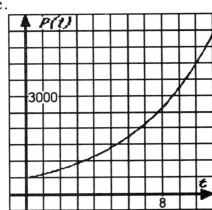

12. a. $P(t) = 8,000\left(\dfrac{1}{2}\right)^{t/5}$

     b. $P(10) = 8,000\left(\dfrac{1}{2}\right)^2 = 2,000$

           $P(28) = 8,000\left(\dfrac{1}{2}\right)^{5.6} = 164$

     c.
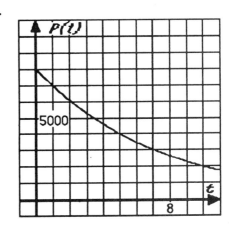

14. a. $P(t) = 15,000(0.9)^{t/3}$

     b. $P(9) = 15,000(0.9)^3 = 10,935$
           $P(10) = 15,000(0.9)^{10/3} = 10,558$

     c.
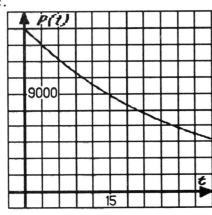

16. a. $P(t) = 12(0.917)^t$

     b. $P(7) = 12(0.917)^7 = 6.54$ g.
           $P(15) = 12(0.917)^{15} = 3.27$ g.

     c.
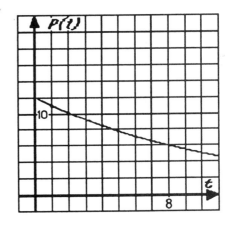

18. For strain A, $P(t) = 0.01(1.1)^{t/8} = 0.01(1.1^{1/8})^t$, while for strain B, $P(t) = 0.01(1.12)^{t/9} = 0.01(1.12^{1/9})^t$. Thus the growth factor for A is $1.1^{1/8}$ or 1.012 and for B is $1.12^{1/9}$ or 1.013, so strain B grows faster.

20. a. $P(t) = 5 + \frac{4}{3}t$ b. $P(t) = 5(1.8)^{t/3}$

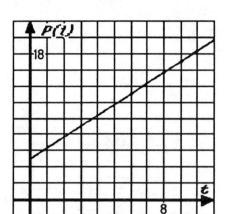 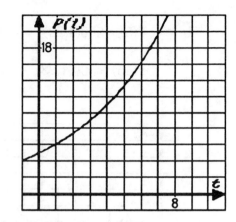

22. $P(x) = 1 \cdot 2^{x-1}$ or $P(x) = 0.5 \cdot 2^x$ $P(64) = 2^{63} = 9.22 \times 10^{18}$ grains

24. a. A level 1 manager makes at least $6(25) = \$150$; level 2 makes at least $6(0.3)(150) = \$270$; level 3 makes at least $6(0.3)270 = \$486$; level 4 makes at least $6(0.3)486 = \$874.80$; level 5, $\$1,574.64$; level 6, $\$2,834.35$. In general, $P(x) = 150(1.8)^{x-1}$.

     b. $6 + 36 + 216 + 1,296 + 7,776 = 9,330$ salespersons

     c. It would be almost impossible to recruit so many salespersons.

26. $P(0) = 4,951,600$ and $P(10) = 6,789,400 = 4,951,600r^{10}$. Thus $r^{10} = \dfrac{6,789,000}{4,951,000} =$ 1.37115, so $r = \sqrt[10]{1.37115} = 1.03207$, or a growth rate of 3.21%.

28. a. $P(7) = 3P(0) = P(0)r^7$, so $3 = r^7$, $r = \sqrt[7]{3} = 1.1699$, and the growth rate = 16.99%.

     b. Since $P(0) = 15,000$ and $P(7) = 45,000 = 3P(0)$, the answer is also 16.99%.

     c. No   d. 16.99%

Exercise 8.2

2. 104.561   4. 3,924.814   6. -1.389   8. 0.0385

10.   19.093

12.   -236.640

14.

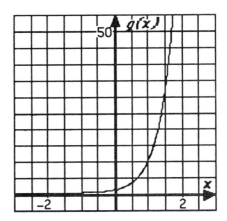

16.

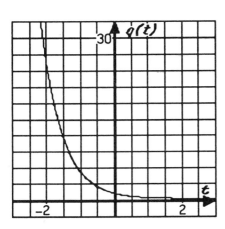

18.

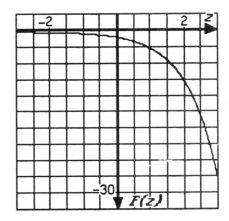

20.

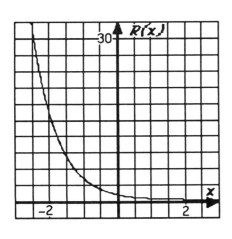

22.

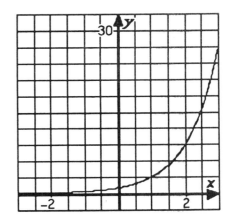

24.

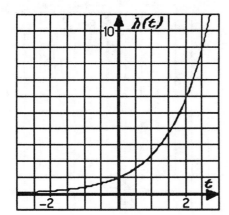

26.

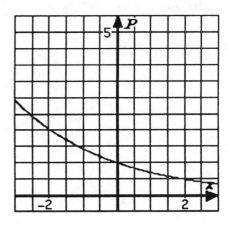

28.

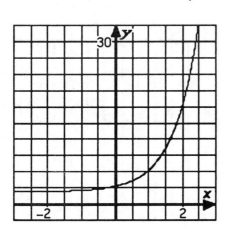

30.

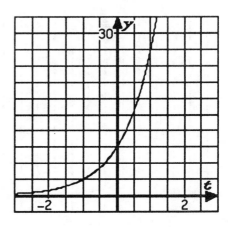

32.

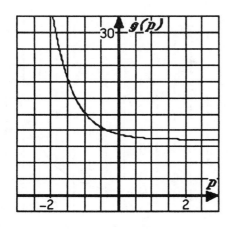

34.

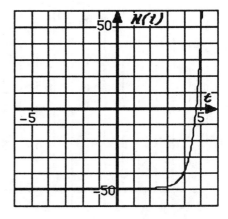

36. $5^x = 125$

$5^x = 5^3$

$x = 3$

38. $3^{x-1} = 27^{1/2}$

$3^{x-1} = \left(3^3\right)^{1/2} = 3^{3/2}$

$x - 1 = \dfrac{3}{2}$

$x = \dfrac{5}{2}$

40. $2^{3x-1} = \dfrac{\sqrt{2}}{16}$

$2^{3x-1} = \dfrac{2^{1/2}}{2^4} = 2^{-7/2}$

$3x - 1 = \dfrac{-7}{2}$

$x = \dfrac{-5}{6}$

120

42. $9 \cdot 3^{x+2} = 81^{-x}$

$3^2 \cdot 3^{x+2} = \left(3^4\right)^{-x}$

$3^{x+4} = 3^{-4x}$

$x + 4 = -4x$

$5x = -4$

$x = \dfrac{-4}{5}$

44. $16^{2-3x} = 64^{x+5}$

$\left(4^2\right)^{2-3x} = \left(4^3\right)^{x+5}$

$4^{4-6x} = 4^{3x+15}$

$4 - 6x = 3x + 15$

$-11 = 9x$

$x = \dfrac{-11}{9}$

46. $5^{x^2-x-4} = 25$

$5^{x^2-x-4} = 5^2$

$x^2 - x - 4 = 2$

$x^2 - x - 6 = 0$

$(x - 3)(x + 2) = 0$

$x = 3 \text{ or } x = -2$

48. $N(t) = 20 \cdot 9^{t/5}$

$14{,}580 = 20 \cdot 9^{t/5}$

$729 = 9^{t/5}$

$9^3 = 9^{t/5}$

$3 = \dfrac{t}{5}$

$t = 15$

It will take 15 weeks

50. $U(t) = 4 \cdot 3^{t/4}$

$324 = 4 \cdot 3^{t/4}$

$81 = 3^{t/4}$

$3^4 = 3^{t/4}$

$4 = \dfrac{t}{4}$

$t = 16$

16 days

52. $V(t) = 20{,}000(0.8)^{t/3}$

$12{,}800 = 20{,}000(0.8)^{t/3}$

$\dfrac{16}{25} = \left(\dfrac{4}{5}\right)^{t/3}$

$\left(\dfrac{4}{5}\right)^2 = \left(\dfrac{4}{5}\right)^{t/3}$

$2 = \dfrac{t}{3}$

$t = 6$

6 years

54. b. and c.

b., because each successive entry doubles, and c., because each successive entry is divided by 10. No such pattern occurs in a. or d.

56.

x	$f(x) = x^3$	$f(x) = 3^x$
-2	-8	1/9
-1	-1	1/3
0	0	1
1	1	3
2	8	9
3	27	27
4	64	81
5	125	243
6	216	729

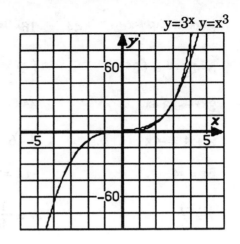

$y=3^x \quad y=x^3$

Exercise 8.3

2.   5   4.   3   6. $\frac{1}{2}$ 8.   -1   10.   1

12.   0   14.   6   16.   -6   18.   3   20.   -3

22. $9^y = 729$ 24. $b^{-3} = 8$ 26. $10^{-4.5} = C$ 28. $5^{6-2p} = 3$

30. $m^p = n$ 32. $\log_{64} \frac{1}{2} = \frac{-1}{6}$ 34. $\log_v 12 = \frac{5}{3}$

36. $\log_{3.7} Q = 2.5$ 38. $\log_z (2P + 5) = -3t$ 40. $\log_{10} 3M_0 = 1.3t$

42. $b^4 = 625 = 5^4$, so b = 5   44. $\left(\frac{1}{2}\right)^5 = x$, so $x = \frac{1}{32}$

46. $5^y = \frac{1}{5} = 5^{-1}$, so y = -1   48. $b^{-1} = 0.1 = \frac{1}{10} = 10^{-1}$, so b = 10

50. $5^3 = 9 - 4x$, so $4x = 9 - 125 = -116$ and x = -29

52. $5(\log_2 x) + 6 = -14$ means $\log_2 x = -4$, so $x = 2^{-4} = \frac{1}{16}$

54. $3^4 = 81$ and $3^5 = 243$, so $4 < \log_3 100 < 5$

56. $10^0 = 1$ and $10^1 = 10$, so $0 < \log_{10} 7 < 1$

58. $6^1 = 6$ and $6^2 = 36$, so $1 < \log_6 24 < 2$

60. $5^2 = 25$ and $5^3 = 125$, so $2 < \log_5 86.3 < 3$

62.   1.4456   64.   3.1691   66.   -2.2118   68.   -1.9830

70. $10^x = 6$ means $\log_{10} 6 = x$, so x = 0.7782

72. $10^{-5x} = 76$ means $\log_{10} 76 = -5x$, so $-5x = 1.8808$ and $x = -0.3762$

74. $8 \cdot 10^{1.6x} = 312$ means $10^{1.6x} = 39$, so $\log_{10} 39 = 1.6x$, $1.6x = 1.5911$, and $x = 0.9944$

76. $163 = 3\left(10^{0.7x}\right) - 49.3$ means $3\left(10^{0.7x}\right) = 212.3$, so $10^{0.7x} = 70.767$ and $0.7x = \log_{10} 70.767$. Thus $0.7x = 1.8498$ and $x = 2.6426$.

78. $4\left(10^{-0.6x}\right) + 16.1 = 28.2$ means $10^{-0.6x} = 3.025$ and $\log_{10} 3.025 = -0.6x$. Thus $-0.6x = 0.4807$ and $x = -0.8012$.

80. $250\left(1 - 10^{-0.3x}\right) = 100$ means $1 - 10^{-0.3x} = 0.4$ and $10^{-0.3x} = 0.6$. Thus $\log_{10} 0.6 = -0.3x$, $-0.3x = -0.2218$, and $x = 0.7393$.

82. $P(20,320 \text{ ft}) = P(3.848 \text{ mi}) = 30 \cdot 10^{-0.09(3.848)} = 30 \cdot 10^{-0.346} = 30(0.4504) = 13.51 \text{ in}$

84. $16.1 = P(a) = 30 \cdot 10^{-0.09a}$, so $10^{-0.09a} = 0.5367$, and $\log_{10} 0.5367 = -0.09a$. Thus $-0.09a = -0.2703$ and $a = 3.003$ miles or $15,856$ feet.

86. At sea level, $a = 0$ and $P(a) = 30(10^{\circ}) = 30$. Thus we seek the value of a for which $a = \frac{1}{4}(30) = 7.5$. Then $7.5 = 30\left(10^{-0.09a}\right)$ or $10^{-0.09a} = 0.25$. Thus $\log_{10} 0.25 = -0.09a$, $-0.09a = -0.60206$, and $a = 6.69$ miles or $35,320$ feet.

88. a. $P(10) = 16,782,000 \cdot 10^{0.036} = 16,782,000(1.086) = 18,232,395$

   b. $P(20) = 16,782,000 \cdot 10^{0.072} = 19,808,141$; $P(30) = 21,520,072$; $P(40) = 23,379,957$

90. $20,000,000 = 16,782,000 \cdot 10^{0.0036t}$, so $10^{0.0036t} = 1.1918$ and $\log_{10} 1.1918 = 0.0036t$. Thus $0.0036t = 0.07619$ and the population reached $20,000,000$ at $t = 21.16$ years or around February, 1981. Repeating with $30,000,000$ yields $t = 70.08$ or early January, 2030.

92. $\log_5 (\log_5 5) = \log_5 1 = 0$

94. $\log_{10} [\log_2 (\log_3 9)] = \log_{10} [\log_2 2] = \log_{10} 1 = 0$

96. $\log_4 [\log_2 (\log_3 81)] = \log_4 [\log_2 4] = \log_4 2 = \log_4 \sqrt{4} = \log_4 4^{1/2} = \frac{1}{2}$

98. $\log_b (\log_a a^b) = \log_b b = 1$

Exercise 8.4

2.

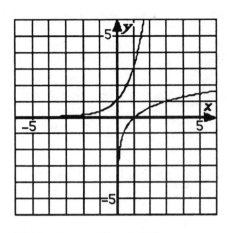

4.

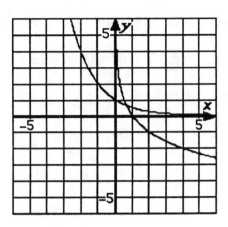

6. $f(93) = \log_{10} 93 = 1.968$ 8. $f(6.95) = \log_{10} 6.95 = 0.842$

10. $f(0)$ is undefined, since the domain is all positive real numbers

12. $3f(41) = 3(\log_{10} 41) = 3(1.6127) = 4.8381$

14. $15 - 4f(7) = 15 - 4(\log_{10} 7) = 15 - 4(0.8451) = 15 - 3.3804 = 11.6196$

16. $\dfrac{3}{2 + f(0.2)} = \dfrac{3}{2 + \log_{10} 0.2} = \dfrac{3}{2 + (-0.6990)} = \dfrac{3}{1.3010} = 2.3059$

18. $k = \dfrac{\log_{10} 35{,}000 - \log_{10} 18{,}000}{26} = \dfrac{4.5441 - 4.2553}{26} = \dfrac{0.2888}{26} = 0.0111$

20. $T = \dfrac{5730 \log_{10} \frac{180}{920}}{\log_{10} \frac{1}{2}} = \dfrac{5730 \log_{10} 0.19565}{-0.3010} = \dfrac{5730(-0.7085)}{-0.3010} = 13{,}487$

22. $h = 56.2 - \sqrt{\dfrac{\log_{10} 78}{0.3}} = 56.2 - \sqrt{\dfrac{1.8921}{0.3}} = 56.2 - \sqrt{6.307} = 56.2 - 2.511 = 53.7$

24.   Solve $\log_{10} x = 2.3$. Then $x = 10^{2.3} = 199.5$.

26.   Solve $\log_{10} x = 0.8$. Then $x = 10^{0.8} = 6.31$.

28.   Solve $\log_{10} x = -1.69$. Then $x = 10^{-1.69} = 0.0204$.

30.   The pH of spinach $= -\log_{10} 3.2 \times 10^{-6} = 5.5$

32. $9.8 = $ pH of ammonia $= -\log_{10} x$. Then $\log_{10} x = -9.8$ and $x = 10^{-9.8}$. Thus the hydrogen ion concentration of ammonia $= 10^{-9.8} = 1.6 \times 10^{-10}$.

124

34. At 100 ft., the decibel level for a jet airplane $= 10 \log_{10}\left(\frac{100}{10^{-12}}\right) = 10 \log_{10} 10^{14} =$ 140 decibels.

36. $10 \log_{10}\left(\frac{I}{10^{-12}}\right) = 210$, so $\log_{10}\left(\frac{I}{10^{-12}}\right) = 21$ and $\frac{I}{10^{-12}} = 10^{21}$. Thus $10^{12}I = 10^{21}$,

    $I = 10^9$, and the intensity is one billion watts per square meter.

38. Let I_s be the intensity of shouting and I_c be the intensity of normal

    conversation. From Example 6 we have $I_c = 10^{-8}$, and from the definition we

    have $123.2 = 10 \log_{10}\left(\frac{I_s}{10^{-12}}\right)$, so $\frac{I_s}{10^{-12}} = 10^{12.32} = 2.09 \times 10^{12}$, so $10^{12}I_s = 2.09 \times 10^{12}$

    and $I_s = 2.09$. Thus $\frac{I_s}{I_c} = \frac{2.09}{10^{-8}} = 2.09 \times 10^8$, and the loudest shout is about 200

    million times more intense than normal conversation.

40. $7.0 = \log_{10}\frac{A_1}{A_0}$; $8.1 = \log_{10}\frac{A_2}{A_0}$. Thus $\frac{A_1}{A_0} = 10^7$, and $\frac{A_2}{A_0} = 10^{8.1}$. Thus $A_1 = 10^7 A_0$

    and $A_2 = 10^{8.1}A_0$, so $\frac{A_2}{A_1} = \frac{10^{8.1}A_0}{10^7 A_0} = 10^{1.1} = 12.6$, and the September quake was

    12.6 times stronger than the April one.

42. Let A be the amplitude of a 3.0 quake. Then $3.0 = \log_{10}\frac{A}{A_0}$, so $\frac{A}{A_0} = 10^3$ and $A = 1000A_0$. A quake 200 times stronger will have amplitude $200A = 200(1000A_0) = 200{,}000A_0$. Its magnitude is given by $\log_{10}\frac{200{,}000A_0}{A_0} = \log_{10} 200{,}000 = 5.3$.

44. a. $f(32) = \log_2 32 = 5$ b. $g[f(32)] = g(5) = 2^5 = 32$

    c. $2^{\log_2 x} = x$ for $x > 0$ because 2^x and $\log_2 x$ are inverse functions and the domain for $\log_2 x$ is all values of x such that $x > 0$.

    d. $2^{\log_2 6}$ must be 6, and $2^{\log_2 Q} = Q$.

46. "Raise 5 to the x power"

48. We want $4^x = 32$, or $2^{2x} = 2^5$, so $2x = 5$ and the answer is $\frac{5}{2}$.

50. We seek an x-value such that $\log_2 x = 10$, that is, $2^{10} = x$, so the solution is 1,024.

125

52. For $\log_b (16 - 3x)$ to be defined, $16 - 3x$ must be positive. Thus it is defined if $16 - 3x > 0$, $16 > 3x$, and $x < \dfrac{16}{3}$. The expression is defined for all values of x less than $\dfrac{16}{3}$.

54. a.

x	$\dfrac{1}{x}$	$\log_{10} x$	$\log_{10} \dfrac{1}{x}$
1	1	0	0
2	$\dfrac{1}{2}$	0.301	-0.301
3	$\dfrac{1}{3}$	0.477	-0.477
4	$\dfrac{1}{4}$	0.602	-0.602
5	$\dfrac{1}{5}$	0.699	-0.699
6	$\dfrac{1}{6}$	0.778	-0.778

56.

x	$y = \log_f x$
1	0
2	0.431
4	0.862
16	1.724
$\dfrac{1}{2}$	-0.431
$\dfrac{1}{4}$	-0.862
$\dfrac{1}{16}$	-1.724

b. $\log_{10} \dfrac{1}{x} = -\log_{10} x$

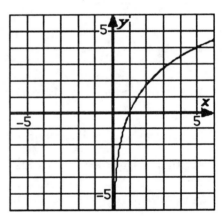

Exercise 8.5

2. $\log_b xy = \log_b x + \log_b y$

4. $\log_b \dfrac{y}{x} = \log_b y - \log_b x$

6. $\log_b \dfrac{x}{yz} = \log_b x - \log_b yz = \log_b x - (\log_b y + \log_b z)$ or $\log_b x - \log_b y - \log_b z$

8. $\log_b x^{1/3} = \dfrac{1}{3} \log_b x$

10. $\log_b \sqrt[5]{y} = \log_b y^{1/5} = \dfrac{1}{5} \log_b y$

12. $\log_b \sqrt{x^3} = \log_b \left(x^3\right)^{1/2} = \log_b x^{3/2} = \dfrac{3}{2} \log_b x$

14. $\log_b x^{1/3} y^2 = \log_b x^{1/3} + \log_b y^2 = \frac{1}{3} \log_b x + 2 \log_b y$

16. $\log_b \frac{xy^2}{z^{1/2}} = \log_b xy^2 - \log_b z^{1/2} = \log_b x + \log_b y^2 - \log_b z^{1/2} =$

$\log_b x + 2 \log_b y - \frac{1}{2} \log_b z$

18. $\log_{10} \sqrt{\frac{2L}{R^2}} = \log_{10} \left(\frac{2L}{R^2}\right)^{1/2} = \log_{10} \frac{(2L)^{1/2}}{R} = \log_{10} (2L)^{1/2} - \log_{10} R =$

$\frac{1}{2} \log_{10} (2L) - \log_{10} R$ or $\frac{1}{2} \log_{10} 2 + \frac{1}{2} \log_{10} L - \log_{10} R$

20. $\log_{10} 2y \sqrt[3]{\frac{x}{y}} = \log_{10} (2y)\frac{x^{1/3}}{y^{1/3}} = \log_{10} 2 + \log_{10} y + \frac{1}{3} \log_{10} x - \frac{1}{3} \log_{10} y =$

$\log_{10} 2 + \frac{1}{3} \log_{10} x + \frac{2}{3} \log_{10} y$

22. $\log_{10} \sqrt{s^2(s-a)^3} = \log_{10} s(s-a)^{3/2} = \log_{10} s + \log_{10} (s-a)^{3/2} =$

$\log_{10} s + \frac{3}{2} \log_{10} (s-a)$

24. $\log_b 10 = \log_b 2(5) = \log_b 2 + \log_b 5 = 0.6931 + 1.6094 = 2.3025$

26. $\log_b \frac{3}{2} = \log_b 3 - \log_b 2 = 1.0986 - 0.6931 = 0.4055$

28. $\log_b 25 = \log_b 5^2 = 2 \log_b 5 = 2(1.6094) = 3.2188$

30. $\log_b \frac{6}{5} = \log_b \frac{2(3)}{5} = \log_b 2 + \log_b 3 - \log_b 5 = 0.6931 + 1.0986 - 1.6094 = 0.1823$

32. $\log_b \sqrt{50} = \log_b (2 \cdot 5^2)^{1/2} = \frac{1}{2} (\log_b 2 + 2 \log_b 5) = \frac{1}{2}(0.6931 + 2(1.6094)) = 1.95595$

34. $\log_b \frac{0.08}{15} = \log_b \frac{8}{1500} = \log_b \frac{2}{375} = \log_b \frac{2}{3 \cdot 5^3} = \log_b 2 - \log_b 3 - 3 \log_b 5 =$

$0.6931 - 1.0986 - 3(1.6094) = -5.2337$

36. $\log_b 5 + \log_b 2 = \log_b 5(2) = \log_b 10$

38. $\frac{1}{4} \log_b x - \frac{3}{4} \log_b y = \log_b x^{1/4} - \log_b y^{3/4} = \log_b \frac{x^{1/4}}{y^{3/4}}$ or $\log_b \sqrt[4]{\frac{x}{y^3}}$

40. $-\log_b x = 0 - \log_b x = \log_b 1 - \log_b x = \log_b \frac{1}{x}$

42. $\frac{1}{3}(\log_{10} x - 2\log_{10} y - \log_{10} z) = \frac{1}{3}(\log_{10} x - \log_{10} y^2 - \log_{10} z) =$

$\frac{1}{3}\left(\log_{10} \frac{x}{y^2 z}\right) = \log_{10}\left(\frac{x}{y^2 z}\right)^{1/3}$ or $\log_{10} \sqrt[3]{\frac{x}{y^2 z}}$

44. $\frac{1}{2}(\log_b 6 + 2\log_b 4) - \log_b 2 = \frac{1}{2}(\log_b 6 + \log_b 4^2) - \log_b 2 = \frac{1}{2}(\log_b 6(16)) - \log_b 2 =$

$\log_b \sqrt{96} - \log_b 2 = \log_b \frac{4\sqrt{6}}{2} = \log_b 2\sqrt{6}$

46. $\log_6 3 + \log_6 x = 2$

$\log_6 3x = 2$

$3x = 6^2 = 36$

$x = 12$

the solution is 12

48. $\log_{10}(x + 3) + \log_{10} x = 1$

$\log_{10} x(x + 3) = 1$

$10^1 = x^2 + 3x$

$x^2 + 3x - 10 = 0$

$(x + 5)(x - 2) = 0$

$x = -5$ does not check, so
the solution is 2

50. $\log_{10}(x - 1) - \log_{10} 4 = 2$

$\log_{10} \frac{x - 1}{4} = 2$

$\frac{x - 1}{4} = 10^2 = 100$

$x - 1 = 400$

$x = 401$

the solution is 401

52. $\log_4(x + 8) + \log_4(x + 2) = 2$

$\log_4(x + 8)(x + 2) = 2$

$x^2 + 10x + 16 = 4^2 = 16$

$x^2 + 10x = 0$

$x(x + 10) = 0$

$x = 0$ or $x = -10$

-10 does not check, so
the solution is 0

128

54. $\log_{10}(x + 3) - \log_{10}(x - 1) = 1$

$\log_{10} \dfrac{x + 3}{x - 1} = 1$

$\dfrac{x + 3}{x - 1} = 10^1 = 10$

$x + 3 = 10x - 10$

$13 = 9x$

$x = \dfrac{13}{9}$

the solution is $\dfrac{13}{9}$

56. $3^x = 4$

$\log_{10} 3^x = \log_{10} 4$

$x \log_{10} 3 = \log_{10} 4$

$x = \dfrac{\log_{10} 4}{\log_{10} 3} = \dfrac{0.6021}{0.4771} = 1.2619$

58. $2^{x-1} = 9$

$\log_{10} 2^{x-1} = \log_{10} 9$

$(x - 1) \log_{10} 2 = \log_{10} 9$

$x - 1 = \dfrac{\log_{10} 9}{\log_{10} 2} = \dfrac{0.9542}{0.3010} = 3.17$

$x = 4.17$

60. $3^{x^2} = 21$

$\log_{10} 3^{x^2} = \log_{10} 21$

$x^2 \log_{10} 3 = \log_{10} 21$

$x^2 = \dfrac{\log_{10} 21}{\log_{10} 3} = \dfrac{1.3222}{0.4771} = 2.7713$

$x = \pm \sqrt{2.7713} = \pm 1.6647$

62. $2.13^{-x} = 8.1$

$\log_{10} 2.13^{-x} = \log_{10} 8.1$

$-x \log_{10} 2.13 = \log_{10} 8.1$

$-x = \dfrac{\log_{10} 8.1}{\log_{10} 2.13} = \dfrac{0.9085}{0.3284} = 2.7765$

$x = -2.7765$

64. $12 \cdot 5^{1.5x} = 85$

$5^{1.5x} = \dfrac{85}{12}$

$\log_{10} 5^{1.5x} = \log_{10} \dfrac{85}{12}$

$1.5x \log_{10} 5 = \log_{10} \dfrac{85}{12}$

$x = \dfrac{\log_{10} \dfrac{85}{12}}{1.5 \log_{10} 5} = \dfrac{0.8502}{1.0485} = 0.8109$

66. $0.06 = 50 \cdot 4^{-0.6x}$

$4^{-0.6x} = \dfrac{0.06}{50} = 0.0012$

$\log_{10} 4^{-0.6x} = \log_{10} 0.0012$

68. Electricity use is given by

$f(t) = \left(4.2 \times 10^6\right)(1.5)^{t/10}$

$10^7 = \left(4.2 \times 10^6\right)(1.5)^{t/10}$

129

$-0.6x \log_{10} 4 = \log_{10} 0.0012$

$$x = \frac{\log_{10} 0.0012}{-0.6 \log_{10} 4} = \frac{-2.9208}{-0.3612} = 8.09$$

$$(1.5)^{t/10} = \frac{10^7}{4.2 \times 10^6} = 2.381$$

$$\log_{10} (1.5)^{t/10} = \log_{10} 2.381$$

$$t = \frac{10 \log_{10} 2.381}{\log_{10} 1.5} = 21.4$$

The demand will be reached in May, 1996 (21.4 years after 1975)

70. An 8000% annual inflation rate means a growth factor of 80, so prices are given by $f(t) = P(80)^t$

$2P = P(80)^t$

$2 = 80^t$

$\log_{10} 2 = t \log_{10} 80$

$$t = \frac{\log_{10} 2}{\log_{10} 80} = \frac{0.3010}{1.9031} = 0.158$$

Prices will double in 0.158 years (less than 2 months)

72. The pollution level is given by

$f(t) = 80(0.9)^t$

$25 = 80(0.9)^t$

$$(0.9)^t = \frac{25}{80} = 0.3125$$

$t \log_{10} (0.9) = \log_{10} 0.3125$

$$t = \frac{\log_{10} 0.3125}{\log_{10} 0.9} = \frac{-0.5052}{-0.04576} = 11.04$$

It will take 11.04 years

74. a. The amount of radium 226 is given by $f(t) = A(0.996)^t$

$$\frac{1}{2} A = A(0.996)^t$$

$0.5 = (0.996)^t$

$t \log_{10} 0.996 = \log_{10} 0.5$

$$t = \frac{\log_{10} 0.5}{\log_{10} 0.996} =$$

76. $B = B_0\left(1 - 10^{-kt}\right)$

$$\frac{B}{B_0} = 1 - 10^{-kt}$$

$$10^{-kt} = 1 - \frac{B}{B_0}$$

$$-kt = \log_{10} \left(1 - \frac{B}{B_0}\right)$$

$$t = \frac{-1}{k} \log_{10} \left(1 - \frac{B}{B_0}\right)$$

$$\frac{-0.3010}{-0.00174} = 172.9$$

The half-life is 172.9 years

b. $\frac{1}{4}A = A(0.996)^t$

$\log_{10} 0.25 = t \log_{10} 0.996$

$t = \frac{\log_{10} 0.25}{\log_{10} 0.996} = 345.9$ years

$\frac{1}{8}A = A(0.996)^t$

$\log_{10} 0.125 = t \log_{10} 0.996$

$t = \frac{\log_{10} 0.125}{\log_{10} 0.996} = 518.8$ years

78. $L = p^a q^b$

$q^b = \frac{L}{p^a}$

$\log_{10} q^b = \log_{10} \frac{L}{p^a}$

$b \log_{10} q = \log_{10} L - \log_{10} p^a$

$b = \frac{\log_{10} L - \log_{10} p^a}{\log_{10} q}$

80. $\log_{10} R = \log_{10} R_0 + kt$

$\log_{10} R - \log_{10} R_0 = kt$

$\log_{10} \frac{R}{R_0} = kt$

$10^{kt} = \frac{R}{R_0}$

$R = R_0 10^{kt}$

82. $\log_b 24 - \log_b 2 = \log_b \frac{24}{2} = \log_b 12 =$

$\log_b 3(4) = \log_b 3 + \log_b 4$

84. $4 \log_b 3 - 2 \log_b 3 = 2 \log_b 3 =$

$\log_b 3^2 = \log_b 9$

86. $\frac{1}{4} \log_b 8 + \frac{1}{4} \log_b 2 = \frac{1}{4}(\log_b 8 + \log_b 2) = \frac{1}{4}(\log_b 8(2)) = \frac{1}{4} \log_b 16 = \log_b 16^{1/4} =$

$\log_b \sqrt[4]{16} = \log_b 2$

88.   Let x = 1000, y = 10. Then $\log_{10} \frac{x}{y} = \log_{10} \frac{1000}{10} = \log_{10} 100 = 2$, but $\frac{\log_{10} x}{\log_{10} y} =$

$\frac{\log_{10} 1000}{\log_{10} 10} = \frac{3}{1} = 3 \neq 2.$

90. Semiannually: $A = 800\left(1 + \dfrac{0.07}{2}\right)^{30} = 800(1.035)^{30} = 800(2.8068) = \$2,245.44$

Monthly: $A = 800\left(1 + \dfrac{0.07}{12}\right)^{180} = 800(1.00583)^{180} = 800(2.8489) = \$2,279.16$

92. Monthly: $1000 = 500\left(1 + \dfrac{0.08}{12}\right)^{12t}$ Daily: $1000 = 500\left(1 + \dfrac{0.08}{365}\right)^{365t}$

$2 = (1.00667)^{12t}$ $2 = (1.0002192)^{365t}$

$\log_{10} 2 = 12t \log_{10} 1.00667$ $\log_{10} 2 = 365t \log_{10} 1.0002192$

$t = \dfrac{\log_{10} 2}{12 \log_{10} 1.00667} = 8.69 \text{ years}$ $t = \dfrac{\log_{10} 2}{365 \log_{10} 1.0002192} = 8.66 \text{ years}$

94. $600 = 400\left(1 + \dfrac{r}{4}\right)^{12}$ 96. $5P = P\left(1 + \dfrac{0.10}{4}\right)^{4t}$

$1.5 = \left(1 + \dfrac{r}{4}\right)^{12}$ $5 = (1.025)^{4t}$

$1 + \dfrac{r}{4} = \sqrt[12]{1.5} = 1.0344$ $\log_{10} 5 = 4t \log_{10} 1.025$

$\dfrac{r}{4} = 0.0344$ $t = \dfrac{\log_{10} 5}{4 \log_{10} 1.025} = 16.3 \text{ years}$

$r = 0.1375 \text{ or } 13.75\%$

Exercise 8.6

2. 1.8405   4.   4.0073   6.   -0.3567   8.   1

10. 2.075   12.   23.57   14.   0.1003   16.   0.5379

18. $e^x = 2.1$ means $x = \ln 2.1 = 0.7419$ 20. $e^x = 60$ means $x = \ln 60 = 4.0943$

22. $e^x = 0.9$ means $x = \ln 0.9 = -0.1054$ 24. $\ln x = 2.03$ means $x = e^{2.03} = 7.614$

26. $\ln x = 0.59$ means $x = e^{0.59} = 1.804$ 28. $\ln x = -3.4$ means $x = e^{-3.4} = 0.0334$

30. $22.26 = 5.3e^{0.4x}$ 32. $14.15 = 4.03e^{1.4x}$ 34. $4.5 = 4e^{2.1x} + 3.3$

$4.2 = e^{0.4x}$ $3.5112 = e^{1.4x}$ $1.2 = 4e^{2.1x}$

$0.4x = \ln 4.2 = 1.4351$ $1.4x = \ln 3.5112 = 1.256$ $0.3 = e^{2.1x}$

$x = 3.588$ $x = 0.897$ $2.1x = \ln 0.3 = -1.204$

                                                    $x = -0.573$

36. $1.23 = 1.3e^{2.1x} - 17.1$ 38. $55.68 = 0.6e^{-0.7x} + 23.1$ 40. $\dfrac{T}{R} = e^{t/2}$

    $18.33 = 1.3e^{2.1x}$ $32.58 = 0.6e^{-0.7x}$ $\dfrac{t}{2} = \ln\dfrac{T}{R}$

    $14.1 = e^{2.1x}$ $54.3 = e^{-0.7x}$ $t = 2\ln\dfrac{T}{R} =$

    $2.1x = \ln 14.1 = 2.6462$ $-0.7x = \ln 54.3 = 3.9945$ $2\ln T - 2\ln R$

    $x = 1.26$ $x = -5.706$

42. $B - 2 = (A + 3)e^{-t/3}$ 44. $P = P_0 + \ln 10k$ 46. $N(t) = 50 \cdot 3^t$

    $\dfrac{B - 2}{A + 3} = e^{-t/3}$ $P - P_0 = \ln 10k$ $3 = e^k$

    $-\dfrac{t}{3} = \ln\dfrac{B - 2}{A + 3}$ $10k = e^{P-P_0}$ $k = \ln 3 = 1.0986$

    $t = -3\ln\dfrac{B - 2}{A + 3}$ $k = \dfrac{1}{10}e^{P-P_0}$ $N(t) = 50e^{1.0986t}$

48. $N(t) = 300(0.8)^t$ 50. $N(t) = 1000(1.04)^t$

    $0.8 = e^k$ $1.04 = e^k$

    $k = \ln 0.8 = -0.223$ $k = \ln 1.04 = 0.0392$

    $N(t) = 300e^{-0.223t}$ $N(t) = 1000e^{0.0392t}$

52.   a. $y = 40e^{-0.4(3)}$ 54.   a. $V = 100\left(1 - e^{-0.5(10)}\right)$

        $y = 40e^{-1.2} = 40(0.301) = 12.05$ g. $V = 100\left(1 - e^{-5}\right) = 99.3$ volts

   b. $12 = 40e^{-0.4t}$ b. $75 = 100\left(1 - e^{-0.5t}\right)$

        $0.3 = e^{-0.4t}$ $0.75 = 1 - e^{-0.5t}$

        $-0.4t = \ln 0.3 = -1.204$ $e^{-0.5t} = 0.25$

        $t = 3.01$ seconds $-0.5t = \ln 0.25 = -1.386$

                                        $t = 2.77$ seconds

56.   a. $A(t) = Pe^{rt}$ 58. $A(t) = Pe^{rt}$

        $A(0.5) = 500e^{0.0625(0.5)}$ $650 = 300e^{0.09t}$

        $A(0.5) = 500e^{0.03125} = \$515.87$ $2.167 = e^{0.09t}$

b. $1200 = Pe^{0.0625(2)}$

$1200 = Pe^{0.125} = P(1.1331)$

$P = \dfrac{1200}{1.1331} = \$1059.04$

$0.09t = \ln 2.167 = 0.773$

$t = 8.59$ years

60. a. $I = I_0 e^{-1.88(1/2)}$

$I = I_0 e^{-0.94} = I_0(0.3906)$

$\dfrac{I}{I_0} = 0.3906$, so 39.1% of the

X-rays penetrate the plate

    b.   To screen out 70%, 30% will pene-

trate, so $I = 0.3I_0 = I_0 e^{-1.88t}$.

Thus $0.3 = e^{-1.88t}$ and $-1.88t =$

$\ln 0.3 = -1.204$, so $t = 0.64$ inches.

62. a. $2 = 1.5e^{k(7)}$

$1.33 = e^{7k}$

$7k = \ln 1.33 = 0.2877$

$k = 0.0411$

$P(t) = 1.5e^{0.0411t}$

    b. $P(19) = 1.5e^{0.0411(19)}$

$P(19) = 1.5e^{0.7809} = \$3.28$

64. $N(t) = N_0\left(\dfrac{1}{2}\right)^{t/12.5}$

$N(t) = N_0\left[\left(\dfrac{1}{2}\right)^{1/12.5}\right]^t = N_0(0.946)^t$

$0.946 = e^k$

$k = \ln 0.946 = -0.0555$

$N(t) = N_0 e^{-0.0555t}$

66. $\log_8 5 = x$

$8^x = 5$

$\log_{10} 8^x = \log_{10} 5$

$x \log_{10} 8 = \log_{10} 5$

$x = \dfrac{\log_{10} 5}{\log_{10} 8} = \dfrac{0.6990}{0.9031} = 0.7740$

68. $\log_b Q = x$

$b^x = Q$

$\log_{10} b^x = \log_{10} Q$

$x \log_{10} b = \log_{10} Q$

$x = \log_b Q = \dfrac{\log_{10} Q}{\log_{10} b}$

70. $\log_{10} Q = x$

$10^x = Q$

$\ln 10^x = \ln Q$

$x \ln 10 = \ln Q$

$x = \log_{10} Q = \dfrac{\ln Q}{\ln 10}$

CHAPTER 9

Exercise 9.1

2. $-2x - 4y = 12$
       $\underline{2x - 3y = 16}$
       $\quad - 7y = 28$

       $y = -4, x = 2$

       the ordered pair (2,-4) is the solution

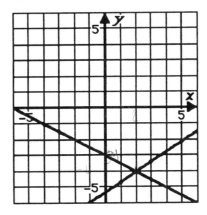

4. $4x - 2y = 14$
       $\underline{3x + 2y = 14}$
       $7x \quad\quad = 28$

       $x = 4, y = 1$

       the ordered pair (4,1) is the solution

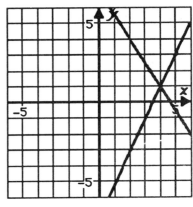

6. $6x + 10y = 2$
       $\underline{-6x + 9y = -21}$
       $\quad\quad 19y = -19$

       $y = -1, x = 2$

       the ordered pair (2,-1) is the solution

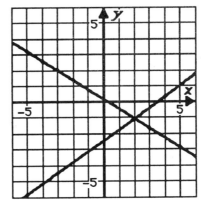

8. $8x - 6 = 6y$
       $\underline{75 + 15x = -6y}$
       $23x + 69 = 0$

       $23x = -69$

       $x = -3, y = -5$

       (-3,-5) is the solution

10. $x - 3y = 7$
       $2x - y = -4$

       $-2x + 6y = -14$
       $\underline{2x - \quad y = -4}$
       $\quad\quad 5y = -18$

       $y = \dfrac{-18}{5}$, $x = \dfrac{-19}{5}$

       $\left(\dfrac{-19}{5}, \dfrac{-18}{5}\right)$ is the solution

12. $x = 2y + 6$
       $2y = x - 6$

       $x - 2y = 6$
       $\underline{-x + 2y = -6}$
       $0x + 0y = 0$

       system is dependent and has

       infinitely many solutions.

135

14. $6x + 8y = 12$
$2x = 12 + 3y$

$6x + 8y = 12$
$\underline{-6x = -36 - 9y}$
$8y = -24 - 9y$

$17y = -24$

$y = \dfrac{-24}{17}, x = \dfrac{66}{17}$

$\left(\dfrac{66}{17}, \dfrac{-24}{17}\right)$ is the solution

16. $64x + 23y = -140.9$
$-52x - 37y = -253.7$

$832x + 299y = -1831.7$
$\underline{-832x - 592y = -4059.2}$
$-293y = -5890.9$

$y = 20.1055, x = -9.427$

$(-9.427, 20.1055)$ is the solution

18. $2x = 60y - 7872$
$11y = 4x + 1083$

$4x - 120y = -15744$
$\underline{-4x + 11y = 1083}$
$-109y = -14661$

$y = 134.5, x = 99.0$

$(99.0, 134.5)$ is the solution

20. $acx + cy = bc$
$\underline{x - cy = d}$
$(ac + 1)x = bc + d$

$x = \dfrac{bc + d}{ac + 1}, y = \dfrac{b - ad}{ac + 1}$

22. $ax + by = c$
$(-1)\underline{ax + by = d}$
$2ax = c + d$

$\searrow 2by = c - d$

$x = \dfrac{c + d}{2a}, y = \dfrac{c - d}{2b}$

24. $-6x + 4y = -12$
$\underline{6x - 4y = 8}$
$0x + 0y = -4$

the system is inconsistent

26. $-12x - 4y = -2$
$\underline{12x + 4y = 2}$
$0x + 0y = 0$

the system is
dependent

28. $6x + 3y = 12$
$\underline{x - 3y = 2}$
$7x = 14$

the system is consis-
tent and independent

30. $-3x - 4y = 8$
$\underline{3x + 4y = -8}$
$0x + 0y = 0$

the system is dependent

32.   first-class passengers: f

tourist passengers: t

$f + t = 42$
$80f + 64t = 2880$

$-64f - 64t = -2688$
$\underline{80f + 64t = 2880}$
$16f = 192$

$f = 12, t = 30$

there were 12 first-class and 30
tourist passengers

34.   true-false points: t

fill-in points: f

$13t + 9f = 71$
$9t + 13f = 83$

$-117t - 81f = -639$
$\underline{117t + 169f = 1079}$
$88f = 440$

$f = \dfrac{440}{88} = 5, t = 2$

fill-ins are worth 5 points and true-
false questions are worth 2 points

36.   8% stock: e

12% stock: t

$e + t = 1200$
$0.08e = 3 + 0.12t$

38.   20% solution: t

15% solution: f

$t + f = 10$
$0.20t + 0.15f = 0.17(10)$

$$12e + 12t = 14400$$
$$\underline{8e - 12t = 300}$$
$$20e = 14700$$

$$e = 735, t = 465$$

$735 at 8%, $465 at 12%

$$-15t - 15f = -150$$
$$\underline{20t + 15f = 170}$$
$$5t = 20$$

$$t = 4, f = 6$$

4 L of 20% solution, 6 L of 15% solution

40. slower car: s

faster car: f

$$f = 2s$$
$$3f = 3s + 96$$

$$-3f = -6s$$

$$\underline{3f = 3s + 96}$$

$$0 = -3s + 96$$

$$3s = 96$$

$$s = 32, f = 64$$

the slower travels 32 mph, the faster travels 64 mph

42. skater's speed: s

wind speed: w

$$(s - w)14.5 = 100$$
$$(s + w)10.5 = 100$$

$$s - w = \frac{100}{14.5}$$

$$s + w = \frac{100}{10.5}$$

$$2s = \frac{100}{14.5} + \frac{100}{10.5} = 16.42$$

$$s = 8.21 \text{ m/sec}, w = 1.314 \text{ m/sec}$$

the skater travels at 8.21 m/sec or 18.58 mph, the wind speed is 1.314 m/sec or 2.97 mph.

44. sports coupe: s

wagon: w

$$3s + 4w = 120$$
$$4s + 5w = 155$$

$$12s + 16w = 480$$
$$\underline{-12s - 15w = -465}$$
$$w = 15$$

$$w = 15, s = 20$$

20 sport coupes, 15 wagons

46. printers: p

terminals: t

$$p = \frac{1}{4}t$$
$$280p + 760t = 19920$$

$$280\left(\frac{1}{4}t\right) + 760t = 19920$$

$$830t = 19920$$

$$t = 24, p = 6$$

24 terminals, 6 printers

48. a. C = 0.6x + 80   b.   R = 1.5x   50.

c.

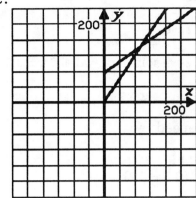

a. and b.

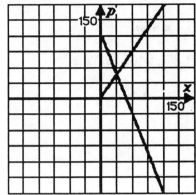

d.   0.6x + 80 = 1.5x

6x + 800 = 15x

9x = 800

x = 88.9

sell 89 loaves to break even

c.   p = 120 - 2.5x
p = 1.5x

6p = 720 - 15x
10p =   15x
16p = 720

p = 45, x = 30

equilibrium price is \$45;
30 pools will be cleaned.

52. a.

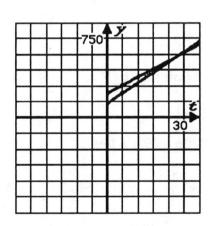

b.   230 + 13t = 140 + 16t

90 = 3t

t = 30

30 months

54.   -6a + b = -7
-3b - a = -17

-6a + b = -7
6a + 18b = 102
19b = 95

b = 5, a = 2

56.   4 = m(-2) + b
-3 = m(5) + b

-4 = 2m - b
-3 = 5m +b
-7 = 7m

m = -1, b = 2

the equation is y = -x + 2

58. u + 2v = $\frac{-11}{12}$

u + v = $\frac{-7}{12}$

60.   u + 2v = 11

u + 2v = -1

62. $\frac{2}{3}$ u + $\frac{3}{4}$ v = $\frac{7}{12}$

4u - $\frac{3}{4}$ v = $\frac{7}{4}$

$u + 2v = \dfrac{-11}{12}$ 　　　　　　$u + 2v = 11$ 　　　　　　$-16u - 18v = -14$

$\dfrac{-u - v = \dfrac{7}{12}}{}$ 　　　　　　$-u - 2v = 1$ 　　　　　　$16u - 3v = 7$

$v = \dfrac{-4}{12} = \dfrac{-1}{3}, y = -3$ 　　$0u + 0v = 12$ 　　　　$-21v = -7$

$u = \dfrac{-3}{12} = \dfrac{-1}{4}, x = -4$ 　　the system is 　　$v = \dfrac{1}{3}, u = \dfrac{1}{2}, x = 2, y = 3$

　　　　　　　　　　　　　inconsistent

(-4,-3) is the solution 　　　　　　　　　　(2,3) is the solution

64. $y = \dfrac{-a_1}{b_1} x + \dfrac{c_1}{b_1}$; $y = \dfrac{-a_2}{b_2} x + \dfrac{c_2}{b_2}$; inconsistent means parallel (same slope, different intercepts), that is, $\dfrac{-a_1}{b_1} = \dfrac{-a_2}{b_2}$ but $\dfrac{c_1}{b_1} \neq \dfrac{c_2}{b_2}$. But $\dfrac{-a_1}{b_1} = \dfrac{-a_2}{b_2}$ means $-a_1b_2 = -a_2b_1$ or $\dfrac{a_1}{a_2} = \dfrac{b_1}{b_2}$. Then $\dfrac{c_1}{b_1} \neq \dfrac{c_2}{b_2}$ means $c_1b_2 \neq b_1c_2$ or $\dfrac{c_1}{c_2} \neq \dfrac{b_1}{b_2}$, as required.

Exercise 9.2

2. $5z = 15$ gives $z = 3$ 　　4. $3x = 3$ gives $x = 1$ 　6. $13x = 13$ gives $x = 1$

$y - 2(3) = -6$ gives $y = 0$ 　　$1 + 4y = 1$ gives $y = 0$ 　$1 - 2z = -7$ gives $z = 4$

$2x + 3(0) - 3 = -7$ gives $x = -2$ 　$1 + 0 + z = 1$ gives $z = 0$ 　$3 - y = 6$ gives $y = -3$

the ordered triple (-2,0,3) is 　the ordered triple 　　the ordered triple
the solution 　　　　　　　(1,0,0) is the solution 　(1,-3,4) is the solution

8. $x - 2y + 4z = -3$ 　　10. $3x - 15y - 3z = 6$ 　12. $3y + z = 3$
$\underline{6x + 2y - 4z = 24}$ 　　　$\underline{3x - 9y + 3z = 6}$ 　　$2x - 3y = -7$
$7x = 21$ 　　　$6x - 24y = 12$ 　$\underline{2x + z = -4}$

$3x + y - 2z = 12$ 　　　$x - 5y - z = 2$ 　　$-4x - 2z = 8$
$\underline{-2x - y + 3z = -11}$ 　　$\underline{-x + 3y + z = 6}$ 　　$\underline{3x + 2z = -6}$
$x + z = 1$ 　　　$-2y = 8$ 　　$-x = 2$

$x = 3, 3 + z = 1$ 　　　$y = -4, 6x + 96 = 12$ 　$x = -2, -6 + 2z = -6$

(3,-1,-2) is the solution 　(-14,-4,4) is the solution 　(-2,1,0) is the solution

14. $-6x + 8y - 4z = -40$ 　16. $3x + 4y + 6z = 2$ 　18. $4x + 8y + 2z = 0$
$\underline{6x - 15y + 15z = 72}$ 　　$\underline{-4x + 4y - 6z = 2}$ 　　$\underline{5x + 3y - 2z = 1}$
$-7y + 11z = 32$ 　　　$-x + 8y = 4$ 　　$9x + 11y = 1$

$4x + 3y - 3z = -4$ 　　　$-6x + 6y - 9z = 3$ 　　$14x + 28y + 7z = 0$
$\underline{-4x + 10y - 10z = -48}$ 　　$\underline{4x - 10y + 9z = 0}$ 　　$\underline{4x - 7y - 7z = 6}$
$13y - 13z = -52$ 　　　$-2x - 4y = 3$ 　　$18x + 21y = 6$

$7y - 7z = -28$ 　　　$-4x - 8y = 6$ 　　　$-18x - 21y = -6$
$\underline{-7y + 11z = 32}$ 　　　$\underline{-x + 8y = 4}$ 　　　$\underline{18x + 22y = 2}$
$4z = 4$ 　　　　　$-5x = 10$ 　　　$y = -4$

139

$z = 1, y - 1 = -4$

$(2,-3,1)$ is the solution

$x = -2, 2 + 8y = 4$

$\left(-2, \dfrac{1}{4}, \dfrac{7}{6}\right)$ is the solution

$y = -4, 9x - 44 = 1$

$(5,-4,6)$ is the solution

20.
$$x + y - 2z = 3$$
$$\underline{6x - 2y + 2z = 10}$$
$$7x - y \quad\;\; = 13$$

$$x + y - 2z = 3$$
$$\underline{-x + y + 2z = -3}$$
$$2y \qquad = 0$$

$y = 0, 7x - 0 = 13$

$\left(\dfrac{13}{7}, 0, \dfrac{-4}{7}\right)$ is the solution

22.
$$x = \left(z + \dfrac{5}{4}\right) + \dfrac{1}{2}$$
$$2z = x - \dfrac{7}{4}$$
$$2z = z + \dfrac{7}{4} - \dfrac{7}{4} = z$$

$z = 0, x = \dfrac{7}{4}$

$\left(\dfrac{7}{4}, \dfrac{5}{4}, 0\right)$ is the solution

24.
$$2x + 6y - 2z = 8$$
$$\underline{-2x - 6y + 2z = 1}$$
$$0x + 0y + 0z = 9$$

the system is incon-
sistent and has no
solution

26.
$$3x + 6y + 2z = -2$$
$$\underline{x - 6y - 2z = 2}$$
$$4x \qquad\quad = 0, x = 0$$

But substituting $x = 0$
reduces all three equations
to $3y + z = -1$, so the system
is dependent

28.
$$x - 2y + z = 5$$
$$\underline{-x + y \quad\;\; = -2}$$
$$-y + z = 3$$
$$\underline{y - z = -3}$$
$$0x + 0y + 0z = 0$$

the system is
dependent

30.
$$x = y + z$$
$$z = 3x - y$$
$$x + z = 3x + z$$

$$-2x = 0, x = 0$$

Substituting $x = 0$ reduces all
three equations to $y + z = 0$,
so the system is dependent

32.
$$2x = y - z + \dfrac{1}{2}$$
$$\underline{-2x = -y + z - 2}$$
$$0x = 0y + 0z - \dfrac{3}{2}$$

the system is inconsistent
and has no solution

34.
$$x + y \qquad = 1$$
$$\underline{2x - y + z = -1}$$
$$3x \qquad + z = 0$$

$$x - 3y - z = \dfrac{-2}{3}$$
$$\underline{3x + 3y \qquad = 3}$$
$$4x \qquad - z = \dfrac{7}{3}$$
$$\underline{3x \qquad + z = 0}$$
$$7x \qquad\quad = \dfrac{7}{3}$$

$x = \dfrac{1}{3}, 1 + z = 0$

$\left(\dfrac{1}{3}, \dfrac{2}{3}, -1\right)$ is the solution

36. 10-dollar bills: t
5-dollar bills: f
1-dollar bills: u

$$t + f + u = 94$$
$$10t + 5f + u = 446$$
$$f = 10 + t$$

$$-t - f - u = -94$$
$$\underline{10t + 5f + u = 446}$$
$$9t + 4f \qquad = 352$$
$$\underline{4t - 4f \qquad = -40}$$
$$13t \qquad\quad = 312$$
$$t = 24, f = 10 + 24$$

140

He had 24 tens, 34
fives, and 36 ones

38. first angle: x
second angle: y
third angle: z

$x + y + z = 180$

$x = 10 + y$

$z = 6y + 10$

$(10 + y) + y + (6y + 10) = 180$

$8y + 20 = 180$

$8y = 160$

$y = 20, x = 30, z = 6(20) + 10 = 130$

the angles are 20°, 30°, and 130°

40. Colombian: c
French roast: f
Sumatran: s

$c + f + s = 1$

$6c + 7.60f + 6.80s = 6.60$

$0.02c + 0.04f + 0.01s = 0.0225$

$$\begin{array}{l} -6c - 6f - 6s = -6 \\ \underline{6c + 7.6f + 6.8s = 6.6} \\ \quad\quad 1.6f + 0.8s = 0.6 \end{array}$$

$$\begin{array}{l} -0.02c - 0.02f - 0.02s = -0.02 \\ \underline{0.02c + 0.04f + 0.01s = 0.0225} \\ \quad\quad 0.02f - 0.01s = 0.0025 \end{array}$$

$$\begin{array}{l} 2f + s = 0.75 \\ \underline{2f - s = 0.25} \\ 4f \quad\;\; = 1, \quad f = 0.25 \end{array}$$

$0.5 + s = 0.75, s = 0.25, c = 0.5$

The house brand contains $\frac{1}{2}$ lb.

Colombian, $\frac{1}{4}$ lb. French roast and

$\frac{1}{4}$ lb. Sumatran

42. Number shipped to L.A.: L
Number shipped to Chicago: C
Number shipped to Miami: M

$L + C + M = 1700$

$L = 2C$

$230L + 70C + 160M = 292{,}000$

Substituting L = 2C in the first and
last equations yields

$3C + M = 1700$

$530C + 160M = 292{,}000$

$$\begin{array}{l} -480C - 160M = -272{,}000 \\ \underline{530C + 160M = 292{,}000} \\ \;\; 50C \quad\quad\quad = 20{,}000, \; C = 400 \end{array}$$

44. acres of wheat: w
acres of corn: c
acres of soybeans: s

$w + c + s = 1300$

$6w + 4c + 5s = 6150$

$5w + 2c + 3s = 3800$

$$\begin{array}{l} -5w - 5c - 5s = -6500 \\ \underline{6w + 4c + 5s = 6150} \\ \;\; w - c \quad\quad = -350 \end{array}$$

$$\begin{array}{l} -3w - 3c - 3s = -3900 \\ \underline{5w + 2c + 3s = 3800} \\ \;\; 2w - c \quad\quad = -100 \\ \underline{\;\; -w + c \quad\quad = 350} \\ \;\; w \quad\quad\quad\;\; = 250 \end{array}$$

$250 - c = -350, c = 600, s = 450$

L = 2(400) = 800, M = 500

Reliable ships 800 sedans to L.A., 400 to Chicago, and 500 to Miami.

Plant 250 acres of wheat, 600 of wheat, and 450 of soybeans.

46.

$4u - 2v + w = 4$

$3u - v + 2w = 0$

$-u + 3v - 2w = 0$

$8u - 4v + 2w = 8$
$\underline{-u + 3v - 2w = 0}$
$7u - v \quad\quad = 8$

$3u - v + 2w = 0$
$\underline{-u + 3v - 2w = 0}$
$2u + 2v \quad\quad = 0$
$14u - 2v \quad\quad = 16$
$16u \quad\quad\quad = 16, u = 1$

$2 + 2v = 0, v = -1, w = -2$

$x = \dfrac{1}{u} = 1, y = \dfrac{1}{v} = -1, z = \dfrac{1}{w} = \dfrac{-1}{2}$

$\left(1, -1, \dfrac{-1}{2}\right)$ is the solution

48.

$2u - v - w = -1$

$4u - 2v + w = -5$

$2u + v - 4w = 4$

$2u - v - w = -1$
$\underline{4u - 2v + w = -5}$
$6u - 3v \quad\quad = -6$

$8u - 4v - 4w = -4$
$\underline{-2u - v + 4w = -4}$
$6u - 5v \quad\quad = -8$
$-6u + 3v \quad\quad = 6$
$\quad -2v \quad\quad = -2, v = 1$

$6u - 5 = -8, u = \dfrac{-1}{2}, w = -1$

$x = \dfrac{1}{u} = -2, y = \dfrac{1}{v} = 1, z = \dfrac{1}{w} = -1$

$(-2, 1, -1)$ is the solution

50.

$0a + 4b + 2c = 1$

$-a + 3b + 0c = 1$

$-a + 0b + 2c = 1$

$-a + 3b = 1$
$\underline{\;a - 2c = -1\;}$
$3b - 2c = 0$
$4b + 2c = 1$
$7b \quad\quad = 1, b = \dfrac{1}{7}$

$-a + \dfrac{3}{7} = 1, \dfrac{3}{7} - 2c = 0$

$a = \dfrac{-4}{7}, \; c = \dfrac{3}{14}$

The coefficients are $a = \dfrac{-4}{7}, b = \dfrac{1}{7}$,

and $c = \dfrac{3}{14}$

52.

$2 = a - b + c$

$6 = a + b + c$

$11 = 4a + 2b + c$

$a - b + c = 2$
$\underline{a + b + c = 6}$
$2a \quad\; + 2c = 8$

$2a - 2b + 2c = 4$
$\underline{4a + 2b + c = 11}$
$6a \quad\quad + 3c = 15$
$\underline{-6a \quad\quad - 6c = -24}$
$\quad\quad\quad -3c = -9, \; c = 3$

$2a + 6 = 8, a = 1$

The parabola has equation
$y = x^2 + 2x + 3$

142

Exercise 9.3

2. $3(1) - 4(-2) = 11$ 4. $1(2) - (-1)(-2) = 0$ 6. $20(-2) - 3(-20) = 20$

8. $(-1)(-6) - (-2)(-5) = -4$

10. $D = 3(-2) - 1(-4) = -2$; $D_x = -2(-2) - 0(-4) = 4$; $D_y = 3(0) - 1(-2) = 2$; $x = \dfrac{4}{-2}$, $y = \dfrac{2}{-2}$; the solution is the ordered pair $(-2,-1)$.

12. $D = 2(-2) - 1(-4) = 0$; the system does not have a unique solution.

14. $D = \dfrac{2}{3}\left(\dfrac{-4}{3}\right) - 1(1) = \dfrac{-17}{9}$; $D_x = 1\left(\dfrac{-4}{3}\right) - 1(0) = \dfrac{-4}{3}$; $D_y = \dfrac{2}{3}(0) - 1(1) = -1$;

   $x = \dfrac{\dfrac{-4}{3}}{\dfrac{-17}{9}} = \dfrac{12}{17}$; $y = \dfrac{-1}{\dfrac{-17}{9}} = \dfrac{9}{17}$; the solution is the ordered pair $\left(\dfrac{12}{17}, \dfrac{9}{17}\right)$.

16. $D = \dfrac{-1}{4}$; $D_x = 0$; $D_y = \dfrac{-3}{4}$; $x = \dfrac{0}{\dfrac{-1}{4}} = 0$; $y = \dfrac{\dfrac{-3}{4}}{\dfrac{-1}{4}} = 3$; the solution is $(0,3)$.

18. $D = 3$; $D_x = 12$; $D_y = -4$; $x = \dfrac{12}{3} = 4$; $y = \dfrac{-4}{3}$; the solution is $\left(4, \dfrac{-4}{3}\right)$.

20. $D = -2$; $D_x = a(-1) - b(1) = -a - b$; $D_y = 1(b) - 1(a) = b - a$; $x = \dfrac{-a - b}{-2}$; $y = \dfrac{b - a}{-2}$; the solution is $\left(\dfrac{1}{2}(a + b), \dfrac{1}{2}(a - b)\right)$.

22. $1(0 - 2) - 3(0 - 0) + 1(-2 - 0) = -4$ 24. $2(6 - 0) - 4(-2 - 8) + (-1)(0 - 12) = 64$

26. $1(4 - 6) - 0(0 - 0) + 0(0 - 0) = -2$ 28. $2(20 + 18) - 1(30 - 30) + 4(-9 - 10) = 0$

30. $-0(3 - 2) + 1(2 + 4) - 0(4 - 12) = 6$ 32. $a(12 - 15) - a(6 - 12) + a(5 - 8) = 0$

34. $0(0 - 0) - 0(0 - 0) + x(0 - x^2) = -x^3$ 36. $0(0 - a^2) - a(0 - ab) + b(a^2 - 0) = 2a^2b$

38. $0(0 - b^2) - b(0 - 0) + 0(b^2 - 0) = 0$

40. $D = -129$; $D_x = 0$; $D_y = -258$; $D_z = 0$; $x = \dfrac{0}{-129} = 0$; $y = \dfrac{-258}{-129} = 2$; $z = \dfrac{0}{-129} = 0$;

   the solution is the ordered triple $(0, 2, 0)$.

42. $D = 36$; $D_x = 72$; $D_y = -108$; $D_z = 36$; $x = \dfrac{72}{36} = 2$; $y = \dfrac{-108}{36} = -3$; $z = \dfrac{36}{36} = 1$;

   the solution is the ordered triple $(2, -3, 1)$.

44. $D = 61$; $D_x = -122$; $D_y = 0$; $D_z = 122$; $x = \dfrac{-122}{61} = -2$; $y = \dfrac{0}{61} = 0$; $z = \dfrac{122}{61} = 2$;

the solution is the ordered triple (-2, 0, 2).

46. $D = 0$; the system has no unique solution.

48. $D = 22$; $D_x = -22$; $D_y = 11$; $D_z = 44$; $x = \dfrac{-22}{22} = -1$; $y = \dfrac{11}{22} = \dfrac{1}{2}$; $z = \dfrac{44}{22} = 2$;

the solution is $(-1, \dfrac{1}{2}, 2)$.

50. $D = 8$; $D_x = 4$; $D_y = 0$; $D_z = 8$; $x = \dfrac{4}{8} = \dfrac{1}{2}$; $y = \dfrac{0}{8} = 0$; $z = \dfrac{8}{8} = 1$;

the solution is $\left(\dfrac{1}{2}, 0, 1\right)$.

52. $D = -3$; $D_x = -21$; $D_y = -12$; $D_z = 15$; $x = \dfrac{-21}{-3} = 7$; $y = \dfrac{-12}{-3} = 4$; $z = \dfrac{15}{-3} = -5$;

the solution is (7, 4, -5).

54. $\begin{vmatrix} a_1 & b_1 \\ a_2 & b_2 \end{vmatrix} = a_1b_2 - b_1a_2 = -(b_1a_2 - a_1b_2) = - \begin{vmatrix} a_2 & b_2 \\ a_1 & b_1 \end{vmatrix}$.

56. $\begin{vmatrix} ka_1 & b_1 \\ ka_2 & b_2 \end{vmatrix} = (ka_1)b_2 - (ka_2)b_1 = k(a_1b_2 - b_1a_2) = k \begin{vmatrix} a_1 & b_1 \\ a_2 & b_2 \end{vmatrix}$.

58. $\begin{vmatrix} a_1 & b_1 \\ ka_2 & kb_2 \end{vmatrix} = a_1(kb_2) - b_1(ka_2) = k(a_1b_2 - b_1a_2) = k \begin{vmatrix} a_1 & b_1 \\ a_2 & b_2 \end{vmatrix}$.

60. If $D = D_x = 0$, then $a_1b_2 - a_2b_1 = c_1b_2 - c_2b_1 = 0$, so $a_1b_2 = a_2b_1$ and $c_1b_2 = c_2b_1$. This implies $\dfrac{a_1}{a_2} = \dfrac{b_1}{b_2}$ and $\dfrac{c_1}{c_2} = \dfrac{b_1}{b_2}$, so $\dfrac{a_1}{a_2} = \dfrac{c_1}{c_2}$, $a_1c_2 = a_2c_1$, and $a_1c_2 - a_2c_1 = D_y = 0$.

62. $\begin{vmatrix} 0 & 0 & 0 \\ a & b & c \\ d & e & f \end{vmatrix} = 0(bf - ce) - 0(af - cd) + 0(ae - bd) = 0$. Any determinant with a row of zeroes has value 0.

64. $\begin{vmatrix} 2 & 0 & 1 \\ 4 & 1 & -2 \\ 6 & 1 & 1 \end{vmatrix} = 4$; $2 \begin{vmatrix} 1 & 0 & 1 \\ 2 & 1 & -2 \\ 3 & 1 & 1 \end{vmatrix} = 2(2) = 4$. Factoring a common factor from each element of a column divides the value of the determinant by that factor.

66. $\begin{vmatrix} x & y & 1 \\ 0 & b & 1 \\ 1 & m & 0 \end{vmatrix} = 0$ is equivalent to $x(0 - m) - y(0 - 1) + 1(0 - b) = 0$, which simplifies to $-mx + y - b = 0$ or $y = mx + b$, which is slope-intercept form.

Exercise 9.4

2. $\begin{bmatrix} \frac{1}{2} & 0 & \Big| & \frac{3}{4} \\ -1 & 5 & \Big| & 4 \end{bmatrix}$

4. $\begin{bmatrix} 1 & -4 & \Big| & 8 \\ 0 & 10 & \Big| & -14 \end{bmatrix}$

6. $\begin{bmatrix} 1 & 6 & 0 & \Big| & -2 \\ 0 & 3 & -2 & \Big| & 8 \\ 0 & 0 & 5 & \Big| & -10 \end{bmatrix}$

8. $\begin{bmatrix} 1 & -7 & 5 & \Big| & 2 \\ 0 & 1 & -3 & \Big| & -1 \\ 0 & 0 & -9 & \Big| & 2 \end{bmatrix}$

10. add 2(row 1) to row 2: $\begin{bmatrix} -2 & 3 & \Big| & 0 \\ 0 & 7 & \Big| & 6 \end{bmatrix}$

12. add $\frac{1}{2}$ (row 1) to row 2: $\begin{bmatrix} 6 & 4 & \Big| & -2 \\ 2 & 0 & \Big| & -4 \end{bmatrix}$

14. add 2(row 1) to row 2 and -3(row 1) to row 3: $\begin{bmatrix} 2 & -1 & 3 & \Big| & -1 \\ 0 & -2 & 10 & \Big| & 3 \\ 0 & 5 & -10 & \Big| & 1 \end{bmatrix}$

16. add $\frac{-1}{4}$ (row 1) to row 2 and $\frac{-5}{4}$ (row 1) to row 3: $\begin{bmatrix} 3 & -2 & 4 & \Big| & -4 \\ 1.25 & 3.5 & 0 & \Big| & 3 \\ -4.25 & 3.5 & 0 & \Big| & 4 \end{bmatrix}$

18. add 4(row 1) to row 2, 2(row 1) to row 3, and $\frac{-3}{4}$ (row 2) to row 3: $\begin{bmatrix} -1 & 2 & 3 & \Big| & 3 \\ 0 & 8 & 13 & \Big| & 6 \\ 0 & 0 & \frac{-27}{4} & \Big| & \frac{-1}{2} \end{bmatrix}$

20. -2(row 1) + row 2 $\begin{bmatrix} 1 & -5 & \Big| & 11 \\ 2 & 3 & \Big| & -4 \end{bmatrix} \rightarrow \begin{bmatrix} 1 & -5 & \Big| & 11 \\ 0 & 13 & \Big| & -26 \end{bmatrix}$, $\begin{matrix} x - 5y = 11 \\ 13y = -26 \end{matrix}$, so y = -2 and x - 5(-2)

= 11, so x = 1. The solution is the ordered pair (1, -2).

22. -5(row 1) + row 2 $\begin{bmatrix} 1 & 6 & \Big| & -14 \\ 5 & -3 & \Big| & -4 \end{bmatrix} \rightarrow \begin{bmatrix} 1 & 6 & \Big| & -14 \\ 0 & -33 & \Big| & 66 \end{bmatrix}$, $\begin{matrix} x + 6y = -14 \\ -33y = 66 \end{matrix}$, so y = -2 and

x + 6(-2) = -14, so x = -2. The solution is the ordered pair (-2, -2).

24. exchange rows 1 and 2 $\begin{bmatrix} 3 & -2 & \Big| & 16 \\ 4 & 2 & \Big| & 12 \end{bmatrix} \rightarrow \begin{bmatrix} 4 & 2 & \Big| & 12 \\ 3 & -2 & \Big| & 16 \end{bmatrix} \underset{(1/4)(\text{row } 1)}{\rightarrow} \begin{bmatrix} 1 & 1/2 & \Big| & 3 \\ 3 & -2 & \Big| & 16 \end{bmatrix}$

$\underset{-3(\text{row } 1)+\text{row } 2}{\rightarrow} \begin{bmatrix} 1 & 1/2 & \Big| & 3 \\ 0 & -7/2 & \Big| & 7 \end{bmatrix}$, $\begin{matrix} x + \frac{1}{2}y = 3 \\ \frac{-7}{2}y = 7 \end{matrix}$, so y = -2 and x + $\frac{1}{2}$(-2) = 3, so x = 4.

The solution is the ordered pair (4, -2).

26. $\frac{1}{4}$(row 1) $\begin{bmatrix} 4 & -3 & | & 16 \\ 2 & 1 & | & 8 \end{bmatrix} \rightarrow \begin{bmatrix} 1 & -3/4 & | & 4 \\ 2 & 1 & | & 8 \end{bmatrix} \xrightarrow[-2(\text{row 1})+\text{row 2}]{} \begin{bmatrix} 1 & -3/4 & | & 4 \\ 0 & 5/2 & | & 0 \end{bmatrix}$,

$\begin{aligned} x - \frac{3}{4}y &= 4 \\ \frac{5}{2}y &= 0 \end{aligned}$, so $y = 0$ and $x - \frac{3}{4}(0) = 4$, so $x = 4$. The solution is $(4,0)$.

28. $\begin{bmatrix} 1 & -2 & 3 & | & -11 \\ 2 & 3 & -1 & | & 6 \\ 3 & -1 & -1 & | & 3 \end{bmatrix} \xrightarrow[\substack{-2(\text{row 1})+\text{row 2} \\ -3(\text{row 1})+\text{row 3}}]{} \begin{bmatrix} 1 & -2 & 3 & | & -11 \\ 0 & 7 & -7 & | & 28 \\ 0 & 5 & -10 & | & 36 \end{bmatrix} \xrightarrow[\frac{1}{7}(\text{row 2})]{} \begin{bmatrix} 1 & -2 & 3 & | & -11 \\ 0 & 1 & -1 & | & 4 \\ 0 & 5 & -10 & | & 36 \end{bmatrix}$

$\xrightarrow[-5(\text{row 2})+\text{row 3}]{} \begin{bmatrix} 1 & -2 & 3 & | & -11 \\ 0 & 1 & -1 & | & 4 \\ 0 & 0 & -5 & | & 16 \end{bmatrix}$, $\begin{aligned} x - 2y + 3z &= -11 \\ y - z &= 4 \\ -5z &= 16 \end{aligned}$, so $z = \frac{-16}{5}$, $y = \frac{4}{5}$, and $x = \frac{1}{5}$.

The solution is the ordered triple $\left(\frac{1}{5}, \frac{4}{5}, \frac{-16}{5}\right)$.

30. $\begin{bmatrix} 1 & -2 & -2 & | & 4 \\ 2 & 1 & -3 & | & 7 \\ 1 & -1 & -1 & | & 3 \end{bmatrix} \xrightarrow[\substack{-2(\text{row 1})+\text{row 2} \\ -1(\text{row 1})+\text{row 3}}]{} \begin{bmatrix} 1 & -2 & -2 & | & 4 \\ 0 & 5 & 1 & | & -1 \\ 0 & 1 & 1 & | & -1 \end{bmatrix} \xrightarrow[\text{exchange row 2 and 3}]{}$

$\begin{bmatrix} 1 & -2 & -2 & | & 4 \\ 0 & 1 & 1 & | & -1 \\ 0 & 5 & 1 & | & -1 \end{bmatrix} \xrightarrow[-5(\text{row 2})+\text{row 3}]{} \begin{bmatrix} 1 & -2 & -2 & | & 4 \\ 0 & 1 & 1 & | & -1 \\ 0 & 0 & -4 & | & 4 \end{bmatrix}$, $\begin{aligned} x - 2y - 2z &= 4 \\ y + z &= -1 \\ -4z &= 4 \end{aligned}$,

so $z = -1$, $y = 0$, and $x = 2$. The solution is the ordered triple $(2, 0, -1)$.

32. $\begin{bmatrix} 1 & -2 & -5 & | & 2 \\ 2 & 3 & 1 & | & 11 \\ 3 & -1 & -1 & | & 11 \end{bmatrix} \xrightarrow[\substack{-2(\text{row 1})+\text{row 2} \\ -3(\text{row 1})+\text{row 3}}]{} \begin{bmatrix} 1 & -2 & -5 & | & 2 \\ 0 & 7 & 11 & | & 7 \\ 0 & 5 & 14 & | & 5 \end{bmatrix} \xrightarrow[\frac{1}{7}(\text{row 2})]{} \begin{bmatrix} 1 & -2 & -5 & | & 2 \\ 0 & 1 & 11/7 & | & 1 \\ 0 & 5 & 14 & | & 5 \end{bmatrix}$

$\xrightarrow[-5(\text{row 2})+\text{row 3}]{} \begin{bmatrix} 1 & -2 & -5 & | & 2 \\ 0 & 1 & 11/7 & | & 1 \\ 0 & 0 & 43/7 & | & 0 \end{bmatrix}$, $\begin{aligned} x - 2y - 5z &= 2 \\ y + \frac{11}{7}z &= 1 \\ \frac{43}{7}z &= 0 \end{aligned}$, so $z = 0$, $y = 1$, and $x = 4$. The

solution is the ordered triple $(4, 1, 0)$.

34. $\begin{bmatrix} 3 & 0 & -1 & | & 7 \\ 2 & 1 & 0 & | & 6 \\ 0 & 3 & -1 & | & 7 \end{bmatrix} \xrightarrow[\frac{-2}{3}(\text{row 1})+\text{row 2}]{} \begin{bmatrix} 3 & 0 & -1 & | & 7 \\ 0 & 1 & 2/3 & | & 4/3 \\ 0 & 3 & -1 & | & 7 \end{bmatrix} \xrightarrow[-3(\text{row 2})+\text{row 3}]{}$

$\begin{bmatrix} 3 & 0 & -1 & | & 7 \\ 0 & 1 & 2/3 & | & 4/3 \\ 0 & 0 & -3 & | & 3 \end{bmatrix}$, $\begin{aligned} 3x - z &= 7 \\ y + \frac{2}{3}z &= \frac{4}{3} \\ -3z &= 3 \end{aligned}$, so $z = -1$, $y = 2$, and $x = 2$. The solution is $(2, 2, -1)$.

36.

$$\begin{bmatrix} 1 & 1 & -1 & 1 & \bigm| & 2 \\ 0 & 3 & 2 & -1 & \bigm| & 0 \\ 2 & 0 & -2 & 1 & \bigm| & 6 \\ -1 & 1 & 3 & -2 & \bigm| & -2 \end{bmatrix} \rightarrow \begin{bmatrix} 1 & 1 & -1 & 1 & \bigm| & 2 \\ 0 & 3 & 2 & -1 & \bigm| & 0 \\ 0 & -2 & 0 & -1 & \bigm| & 2 \\ 0 & 2 & 2 & -1 & \bigm| & 0 \end{bmatrix} \rightarrow \begin{bmatrix} 1 & 1 & -1 & 1 & \bigm| & 2 \\ 0 & 3 & 2 & -1 & \bigm| & 0 \\ 0 & 1 & 0 & 0.5 & \bigm| & -1 \\ 0 & 2 & 2 & -1 & \bigm| & 0 \end{bmatrix} \rightarrow \begin{bmatrix} 1 & 1 & -1 & 1 & \bigm| & 2 \\ 0 & 1 & 0 & 0.5 & \bigm| & -1 \\ 0 & 3 & 2 & -1 & \bigm| & 0 \\ 0 & 2 & 2 & -1 & \bigm| & 0 \end{bmatrix} \rightarrow$$

$$\begin{bmatrix} 1 & 1 & -1 & 1 & \bigm| & 2 \\ 0 & 1 & 0 & 0.5 & \bigm| & -1 \\ 0 & 0 & 2 & -2.5 & \bigm| & 3 \\ 0 & 0 & 2 & -2 & \bigm| & 2 \end{bmatrix} \rightarrow \begin{bmatrix} 1 & 1 & -1 & 1 & \bigm| & 2 \\ 0 & 1 & 0 & 0.5 & \bigm| & -1 \\ 0 & 0 & 2 & -2.5 & \bigm| & 3 \\ 0 & 0 & 0 & 0.5 & \bigm| & -1 \end{bmatrix} \rightarrow \begin{bmatrix} 1 & 1 & -1 & 1 & \bigm| & 2 \\ 0 & 2 & 0 & 1 & \bigm| & -2 \\ 0 & 0 & 4 & -5 & \bigm| & 6 \\ 0 & 0 & 0 & 1 & \bigm| & -2 \end{bmatrix}$$

38.

$$\begin{bmatrix} 3 & -2 & 1 & 0 & \bigm| & 4 \\ 1 & 4 & -2 & 1 & \bigm| & -5 \\ 1 & 0 & -1 & 0 & \bigm| & -3 \\ -2 & 2 & 1 & -1 & \bigm| & 1 \end{bmatrix} \rightarrow \begin{bmatrix} 1 & 0 & -1 & 0 & \bigm| & -3 \\ 1 & 4 & -2 & 1 & \bigm| & -5 \\ 3 & -2 & 1 & 0 & \bigm| & 4 \\ -2 & 2 & 1 & -1 & \bigm| & 1 \end{bmatrix} \rightarrow \begin{bmatrix} 1 & 0 & -1 & 0 & \bigm| & -3 \\ 0 & 4 & -1 & 1 & \bigm| & -2 \\ 0 & -2 & 4 & 0 & \bigm| & 13 \\ 0 & 2 & -1 & -1 & \bigm| & -5 \end{bmatrix} \rightarrow \begin{bmatrix} 1 & 0 & -1 & 0 & \bigm| & -3 \\ 0 & -2 & 4 & 0 & \bigm| & 13 \\ 0 & 4 & -1 & 1 & \bigm| & -2 \\ 0 & 2 & -1 & -1 & \bigm| & -5 \end{bmatrix} \rightarrow$$

$$\begin{bmatrix} 1 & 0 & -1 & 0 & \bigm| & -3 \\ 0 & -2 & 4 & 0 & \bigm| & -13 \\ 0 & 0 & 7 & 1 & \bigm| & 24 \\ 0 & 0 & 3 & -1 & \bigm| & 8 \end{bmatrix} \rightarrow \begin{bmatrix} 1 & 0 & -1 & 0 & \bigm| & -3 \\ 0 & -2 & 4 & 0 & \bigm| & -13 \\ 0 & 0 & 1 & 1/7 & \bigm| & 24/7 \\ 0 & 0 & 0 & -10/7 & \bigm| & -16/7 \end{bmatrix} \rightarrow \begin{bmatrix} 1 & 0 & -1 & 0 & \bigm| & -3 \\ 0 & -2 & 4 & 0 & \bigm| & -13 \\ 0 & 0 & 7 & 1 & \bigm| & 24 \\ 0 & 0 & 0 & -10 & \bigm| & -16 \end{bmatrix}$$

40.

$$\begin{bmatrix} 1 & -1 & 1 & -1 & \bigm| & 8 \\ 1 & 3 & 0 & 2 & \bigm| & -1 \\ -2 & -2 & 1 & -1 & \bigm| & -3 \\ 3 & 1 & 0 & 1 & \bigm| & 10 \end{bmatrix} \rightarrow \begin{bmatrix} 1 & -1 & 1 & -1 & \bigm| & 8 \\ 0 & 4 & -1 & 3 & \bigm| & -9 \\ 0 & -4 & 3 & -3 & \bigm| & 13 \\ 0 & 4 & -3 & 4 & \bigm| & -14 \end{bmatrix} \rightarrow \begin{bmatrix} 1 & -1 & 1 & -1 & \bigm| & 8 \\ 0 & 4 & -1 & 3 & \bigm| & -9 \\ 0 & 0 & 2 & 0 & \bigm| & 4 \\ 0 & 0 & -2 & 1 & \bigm| & -5 \end{bmatrix} \rightarrow \begin{bmatrix} 1 & -1 & 1 & -1 & \bigm| & 8 \\ 0 & 4 & -1 & 3 & \bigm| & -9 \\ 0 & 0 & 1 & 0 & \bigm| & 2 \\ 0 & 0 & 0 & 1 & \bigm| & -1 \end{bmatrix},$$

$$\begin{aligned} x - y + z - w &= 8 \\ 4y - z + 3w &= -9 \\ z &= 2 \\ w &= -1 \end{aligned}$$, so $w = -1$, $z = 2$, $y = -1$, and $x = 4$. The solution is $(4, -1, 2, -1)$.

42.

$$\begin{bmatrix} 2 & 1 & 2 & 0 & \bigm| & 3 \\ -2 & -1 & 0 & 3 & \bigm| & -1 \\ 3 & -2 & 1 & 1 & \bigm| & 7 \\ 3 & 1 & 3 & 2 & \bigm| & 20 \end{bmatrix} \rightarrow \begin{bmatrix} 2 & 1 & 2 & 0 & \bigm| & 3 \\ 0 & 0 & 2 & 3 & \bigm| & 2 \\ 3 & -2 & 1 & 1 & \bigm| & 0 \\ 0 & 3 & -4 & 1 & \bigm| & 13 \end{bmatrix} \rightarrow \begin{bmatrix} 2 & 1 & 2 & 0 & \bigm| & 3 \\ 0 & 0 & 2 & 3 & \bigm| & 2 \\ 0 & -7/2 & -2 & 1 & \bigm| & 5/2 \\ 0 & 3 & -4 & 1 & \bigm| & 13 \end{bmatrix} \rightarrow \begin{bmatrix} 2 & 1 & 2 & 0 & \bigm| & 3 \\ 0 & 3 & -4 & 1 & \bigm| & 13 \\ 0 & -7/2 & -2 & 1 & \bigm| & 5/2 \\ 0 & 0 & 2 & 3 & \bigm| & 2 \end{bmatrix} \rightarrow$$

$$\begin{bmatrix} 2 & 1 & 2 & 0 & \bigm| & 3 \\ 0 & 3 & -4 & 1 & \bigm| & 13 \\ 0 & 0 & -20/3 & 13/6 & \bigm| & 53/3 \\ 0 & 0 & 2 & 3 & \bigm| & 2 \end{bmatrix} \rightarrow \begin{bmatrix} 2 & 1 & 2 & 0 & \bigm| & 3 \\ 0 & 3 & -4 & 1 & \bigm| & 13 \\ 0 & 0 & -20 & 13/2 & \bigm| & 53 \\ 0 & 0 & 2 & 3 & \bigm| & 2 \end{bmatrix} \rightarrow \begin{bmatrix} 2 & 1 & 2 & 0 & \bigm| & 3 \\ 0 & 3 & -4 & 1 & \bigm| & 13 \\ 0 & 0 & 2 & 3 & \bigm| & 2 \\ 0 & 0 & -20 & 13/2 & \bigm| & 53 \end{bmatrix} \rightarrow$$

$$\begin{bmatrix} 2 & 1 & 2 & 0 & \bigm| & 3 \\ 0 & 3 & -4 & 1 & \bigm| & 13 \\ 0 & 0 & 2 & 3 & \bigm| & 2 \\ 0 & 0 & 0 & 73/2 & \bigm| & 73 \end{bmatrix},$$

$$\begin{aligned} 2x + y + 2z &= 3 \\ 3y - 4z + w &= 13 \\ 2z + 3w &= 2 \\ \frac{73}{2} w &= 73 \end{aligned}$$, so $w = 2$, $z = -2$, $y = 1$, and $x = 3$. The

solution is $(3, 1, -2, 2)$.

CHAPTER 10

Exercise 10.1

2. $2x + 3 \leq x - 1$

     $x \leq -4$

     $(-\infty, -4]$

4. $-3(-1 - 2x) \geq 5x - 6$

     $3 + 6x \geq 5x - 6$

     $x \geq -9$

     $[-9, +\infty)$

6. $-6 - 2(x - 4) < 5x + 2$

     $-6 - 2x + 8 < 5x + 2$

     $0 < 7x$ $x > 0$

     $(0, +\infty)$

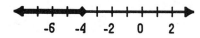 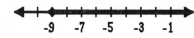 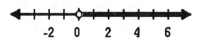

8. $\dfrac{-2x - 3}{2} \leq -5$

     $-2x - 3 \leq -10$

     $-2x \leq -7$

     $x \geq \dfrac{7}{2}$

     $\left[\dfrac{7}{2}, +\infty\right)$

10. $\dfrac{3x - 4}{-2} > \dfrac{-2x}{5}$

     $-15x + 20 > -4x$

     $-11x > -20$

     $x < \dfrac{20}{11}$

     $\left(-\infty, \dfrac{20}{11}\right)$

12. $\dfrac{1}{3}(2x + 5) \leq \dfrac{-5}{6}(x - 2)$

     $4x + 10 \leq -5x + 10$

     $9x \leq 0$

     $x \leq 0$

     $(-\infty, 0]$

 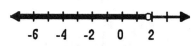 

14. $\dfrac{-3}{2}(2x + 1) - \dfrac{1}{4}(x + 1) < 3$

     $-12x - 6 - x - 1 < 12$

     $-13x < 19$

     $x > \dfrac{-19}{13}$

     $\left(\dfrac{-19}{13}, +\infty\right)$

16. $-3 \leq 3 - 2x < 9$

     $-6 \leq -2x < 6$

     $3 \geq x > -3$

     $(-3, 3]$

18. $0 \geq \dfrac{x + 5}{2} \geq -2$

     $0 \geq x + 5 \geq -4$

     $-5 \geq x \geq -9$

     $[-9, -5]$

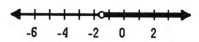  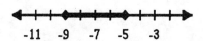

20. $0.2 < \dfrac{2x - 1.4}{4} \le 2.6$

   $0.8 < 2x - 1.4 \le 10.4$

   $2.2 < 2x \le 11.8$

   $1.1 < x \le 5.9$

   $(1.1, 5.9]$

22. $[-1, 0]$

24. $\varnothing$

26. $\varnothing$

28. $(-5, -3)$

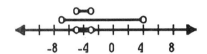

30. $(-\infty, 5] \cap [1, +\infty) = [1, 5]$

32. $2x - 3 < 5$ and $-2x + 3 < 5$

   $2x < 8$ and $-2x < 2$

   $x < 4$ and $x > -1$

   $(-\infty, 4) \cap (-1, +\infty) = (-1, 4)$

34. $9 - 3x < 21$ and $-10 < -5x + 15$

   $-3x < 12$ and $-25 < -5x$

   $x > -4$ and $x < 5$

   $(-4, +\infty) \cap (-\infty, 5) = (-4, 5)$

36. $[-6, 1]$ 38. already simplified   40. $(-6, -2]$

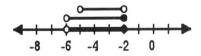

42. already simplified

44. $x - 3 \ge 6$ or $x + 4 < 16$

   $x \ge 9$ or $x < 12$

   $[9, +\infty) \cup (-\infty, 12) = (-\infty, +\infty)$

46. $3x - 1 \le 8$ or $3 - 2x > 7$

   $3x \le 9$ or $-2x > 4$

   $x \le 3$ or $x < -2$

   $(-\infty, 3] \cup (-\infty, -2) = (-\infty, 3]$

149

48. $6x + 2 \geq -16$ or $3x + 8 < -10$

$6x \geq -18$ or $3x < -18$

$x \geq -3$ or $x < -6$

$[-3, +\infty) \cup (-\infty, -6)$

50. $-15 \leq F \leq 25$

$-15 \leq \frac{9}{5} C + 32 \leq 25$

$-47 \leq \frac{9}{5} C \leq -7$

$-26\frac{1}{9} \leq C \leq -3\frac{8}{9}$

The comfort range is -26.1°C to -3.9°C.

52. Income next month: x

$$\frac{16000+17600+19500+18800+22000+x}{6} \geq 20000$$

$$\frac{93900 + x}{6} \geq 20000$$

$93900 + x \geq 120000$

$x \geq 26100$

Owen's Market must make at least $26,100 next month.

54. x: cost of visit

$10 + 0.3(x - 10) < 15 + 0.2(x - 15)$

$100 + 3x - 30 < 150 + 2x - 30$

$70 + 3x < 120 + 2x$

$x < 50$

Plan A is cheaper when a visit costs less than $50.

56. Pace for 20 miles: x

$$6.5 \leq \frac{24(8)+6(5)+20x}{50} \leq 7$$

$325 \leq 222 + 20x \leq 350$

$103 \leq 20x \leq 128$

$5.15 \leq x \leq 6.4$

She must run between 5.15 and 6.4 minutes per mile.

58. Clerk's salary: x

$$\frac{4(28000)+12(22000)+30x}{46} \leq 19000$$

$376000 + 30x \leq 874000$

$30x \leq 498000$

$x \leq 16600$

Lacy's can pay its clerks up to $16,600.

Exercise 10.2

2. $x - 3 = 2$ or $-(x - 3) = 2$

$x = 5$ or $x - 3 = -2$

$x = 5$ or $x = 1$

the solutions are 5 and 1

4. $x + 6 = 1$ or $-(x + 6) = 1$

$x = -5$ or $x + 6 = -1$

$x = -5$ or $x = -7$

the solutions are -5 and -7

6. $3x - 1 = 5$ or $-(3x - 1) = 5$

       $3x = 6$ or $3x - 1 = -5$

       $x = 2$ or $3x = -4$

       $x = 2$ or $x = \dfrac{-4}{3}$

       the solutions are 2 and $\dfrac{-4}{3}$

8. $6 - 5x = 4$ or $-(6 - 5x) = 4$

       $-5x = -2$ or $6 - 5x = -4$

       $x = \dfrac{2}{5}$ or $-5x = -10$

       $x = \dfrac{2}{5}$ or $x = 2$

       the solutions are $\dfrac{2}{5}$ and 2

10. $x^2 - 2x - 4 = 4$ or $-(x^2 - 2x - 4) = 4$

       $x^2 - 2x - 8 = 0$ or $x^2 - 2x = 0$

       $(x - 4)(x + 2) = 0$ or $x(x - 2) = 0$

       $x = 4$ or $x = -2$ or $x = 0$ or $x = 2$

       the solutions are 4, 2, 0, and -2

12. $\left| a \right| = 7$

14. $\left| q - (-7) \right| = 2$ or $\left| q + 7 \right| = 2$

16. $\left| w + 5 \right| \le 1$

18. $\left| m - 8 \right| > 0.1$

20. $-5 < x < 5$

      $(-5, 5)$

22. $-8 \le x + 1 \le 8$

      $-9 \le x \le 7$

      $[-9, 7]$

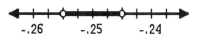

24. $-6 < 2x + 4 < 6$

      $-10 < 2x < 2$

      $-5 < x < 1$ $(-5, 1)$

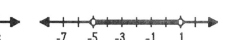

26. $-15 \le 5 - 3x \le 15$

      $-20 \le -3x \le 10$

      $\dfrac{20}{3} \ge x \ge \dfrac{-10}{3}$

      $\left[\dfrac{-10}{3}, \dfrac{20}{3} \right]$

28. $-0.02 < 4x + 1 < 0.02$

      $-1.02 < 4x < -0.98$

      $-0.255 < x < -0.245$

      $(-0.255, -0.245)$

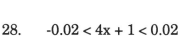

30. $x \ge 5$ or $x \le -5$

      $(-\infty, -5] \cup [5, +\infty)$

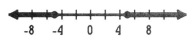

32. $x + 5 > 2$ or $x + 5 < -2$

      $x > -3$ or $x < -7$

34. $4 - 3x > 10$ or $4 - 3x < -10$

      $-3x > 6$ or $-3x < -14$

$(-\infty, -7) \cup (-3, +\infty)$

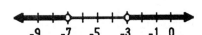

$x < -2 \text{ or } x > \dfrac{14}{3}$

$(-\infty, -2) \cup \left(\dfrac{14}{3}, +\infty\right)$

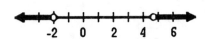

36. $2x - 3.2 \geq 1.4 \text{ or } 2x - 3.2 \leq -1.4$

$2x \geq 4.6 \text{ or } 2x \leq 1.8$

$x \geq 2.3 \text{ or } x \leq 0.9$

$(-\infty, 0.9] \cup [2.3, +\infty)$

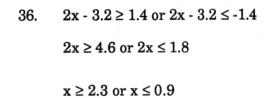

38. $1 - 7x > 0 \text{ or } 1 - 7x < 0$

$-7x > -1 \text{ or } -7x < -1$

$x < \dfrac{1}{7} \text{ or } x > \dfrac{1}{7}$

$\left(-\infty, \dfrac{1}{7}\right) \cup \left(\dfrac{1}{7}, +\infty\right)$

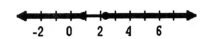

40. $2x - 5 > \dfrac{3}{2} \text{ or } 2x - 5 < \dfrac{-3}{2}$

$2x > \dfrac{13}{2} \text{ or } 2x < \dfrac{7}{2}$

$x > \dfrac{13}{4} \text{ or } x < \dfrac{7}{4}$

$\left(-\infty, \dfrac{7}{4}\right) \cup \left(\dfrac{13}{4}, +\infty\right)$

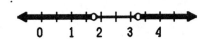

42. $x + \dfrac{1}{2} \geq \dfrac{1}{3} \text{ or } x + \dfrac{1}{2} \leq \dfrac{-1}{3}$

$x \geq \dfrac{-1}{6} \text{ or } x \leq \dfrac{-5}{6}$

$\left(-\infty, \dfrac{-5}{6}\right] \cup \left[\dfrac{-1}{6}, +\infty\right)$

44. $27 - 5x > 12.5 \text{ or } 27 - 5x < -12.5$

$-5x > -14.5 \text{ or } -5x < -39.5$

$x < 2.9 \text{ or } x > 7.9$

$(-\infty, 2.9) \cup (7.9, +\infty)$

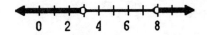

46. $-0.01 < 3x + 2.1 < 0.01$

$-2.11 < 3x < -2.09$

$-0.703 < x < -0.697$

$(-0.703, -0.697)$

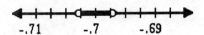

152

48. $|L - 78| \leq 0.5$

50. a. $|C - 56.2| \leq 0.4$

b. $-0.4 \leq 2\pi R - 56.2 \leq 0.4$

$55.8 \leq 2\pi R \leq 56.6$

$\dfrac{27.9}{\pi} \leq R \leq \dfrac{28.3}{\pi}$

$8.88 \leq R \leq 9.008$

$-0.064 \leq R - 8.944 \leq 0.064$

the radius is 8.944 ± 0.064

52. weight of bag: x

5000 lb = 125 40-lb bags

$|5000 - 125x| \leq 5$

$-5 \leq 5000 - 125x \leq 5$

$-5005 \leq -125x \leq -4995$

$39.96 \leq x \leq 40.04$

the bags must weigh between 39.96 and 40.04 lb.

54. Speed in yards per second: x

$|15x - 366| \leq 1$

$-1 \leq 15x - 366 \leq 1$

$365 \leq 15x \leq 367$

$24\dfrac{1}{3} \leq x \leq 24\dfrac{7}{15}$

The cyclist's speed is between

$24\dfrac{1}{3}$ and $24\dfrac{7}{15}$ yards per second, which is between 49.77 and 50.05 mph.

56. $|P - Q| + |P - N| = 6$

58. $|x - d| > |x - 5|$

60. $|x| \leq 4$

62. $|x + 5| < 5$

64. $|x + 5| < 3$

66. $\left|x + \dfrac{3}{2}\right| \leq \dfrac{7}{2}$

Exercise 10.3

2. $(x - 3)(x + 2) > 0$

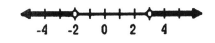

$(-\infty, -2) \cup (3, +\infty)$

4. $(z + 2)(z + 5) \leq 0$

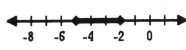

$[-5, -2]$

6. $m(m + 4) > 0$

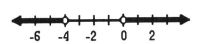

$(-\infty, -4) \cup (0, +\infty)$

8. $p^2 - 5p - 6 \geq 0$

$(p - 6)(p + 1) \geq 0$

10. $2r^2 + r - 10 < 0$

$(2r + 5)(r - 2) < 0$

12. $y^2 - 3y \geq 10$

$y^2 - 3y - 10 \geq 0$

$(y - 5)(y + 2) \geq 0$

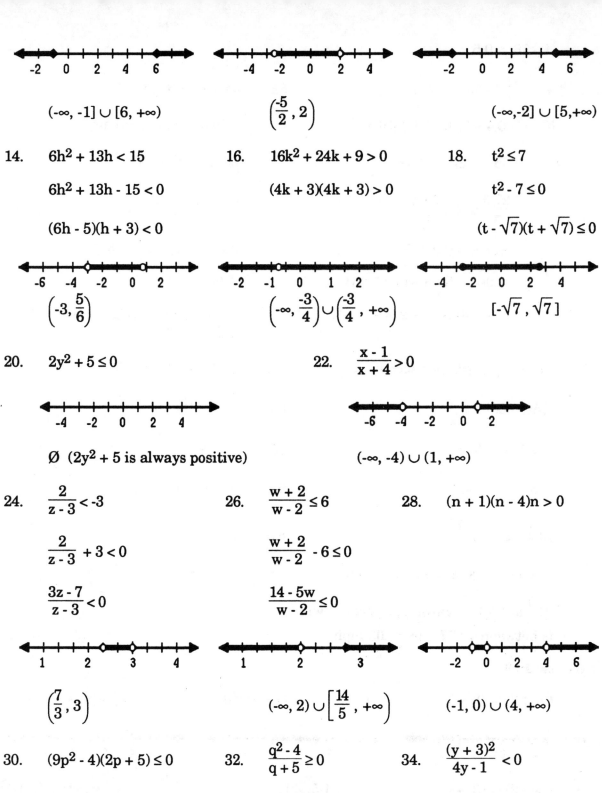

$(-\infty, -1] \cup [6, +\infty)$

$\left(\dfrac{-5}{2}, 2\right)$

$(-\infty,-2] \cup [5,+\infty)$

14. $6h^2 + 13h < 15$

$6h^2 + 13h - 15 < 0$

$(6h - 5)(h + 3) < 0$

16. $16k^2 + 24k + 9 > 0$

$(4k + 3)(4k + 3) > 0$

18. $t^2 \le 7$

$t^2 - 7 \le 0$

$(t - \sqrt{7})(t + \sqrt{7}) \le 0$

$\left(-3, \dfrac{5}{6}\right)$

$\left(-\infty, \dfrac{-3}{4}\right) \cup \left(\dfrac{-3}{4}, +\infty\right)$

$[-\sqrt{7}, \sqrt{7}]$

20. $2y^2 + 5 \le 0$

22. $\dfrac{x - 1}{x + 4} > 0$

Ø ($2y^2 + 5$ is always positive)

$(-\infty, -4) \cup (1, +\infty)$

24. $\dfrac{2}{z - 3} < -3$

$\dfrac{2}{z - 3} + 3 < 0$

$\dfrac{3z - 7}{z - 3} < 0$

26. $\dfrac{w + 2}{w - 2} \le 6$

$\dfrac{w + 2}{w - 2} - 6 \le 0$

$\dfrac{14 - 5w}{w - 2} \le 0$

28. $(n + 1)(n - 4)n > 0$

$\left(\dfrac{7}{3}, 3\right)$

$(-\infty, 2) \cup \left[\dfrac{14}{5}, +\infty\right)$

$(-1, 0) \cup (4, +\infty)$

30. $(9p^2 - 4)(2p + 5) \le 0$

$(3p - 2)(3p + 2)(2p + 5) \le 0$

32. $\dfrac{q^2 - 4}{q + 5} \ge 0$

$\dfrac{(q - 2)(q + 2)}{q + 5} \ge 0$

34. $\dfrac{(y + 3)^2}{4y - 1} < 0$

$\left(-\infty, \dfrac{-5}{2}\right] \cup \left[\dfrac{-2}{3}, \dfrac{2}{3}\right]$

$(-5, -2] \cup [2, +\infty)$

$(-\infty, -3) \cup \left(-3, \dfrac{1}{4}\right)$

36. $\dfrac{b^2 - 16}{3b} \geq -2$

$\dfrac{b^2 - 16}{3b} + 2 \geq 0$

$\dfrac{b^2 + 6b - 16}{3b} \geq 0$

$\dfrac{(b + 8)(b - 2)}{3b} \geq 0$

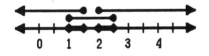

$[-8, 0) \cup [2, +\infty)$

38. $\dfrac{8t}{6t + 1} > 4t$

$\dfrac{8t}{6t + 1} - 4t > 0$

$\dfrac{4t - 24t^2}{6t + 1} > 0$

$\dfrac{4t(1 - 6t)}{6t + 1} > 0$

$\left(-\infty, \dfrac{-1}{6}\right) \cup \left(0, \dfrac{1}{6}\right)$

40. $40 \leq 56t - 16t^2 \leq 48$

$40 \leq 56t - 16t^2$ and $56t - 16t^2 \leq 48$

$16t^2 - 56t + 40 \leq 0$ and $16t^2 - 56t + 48 \geq 0$

$8(2t - 5)(t - 1) \leq 0$ and $8(2t - 3)(t - 2) \geq 0$

$\left[1, \dfrac{3}{2}\right] \cup \left[2, \dfrac{5}{2}\right]$

The ball is between 40 and 48 ft high from 1 to 1.5 sec and from 2 to 2.5 sec.

42. $100x^2 + 4000x + 9000 \leq 185000$

$x^2 + 40x + 90 \leq 1850$

$x^2 + 40x - 1760 \leq 0$

$\left(x - \dfrac{-40 + \sqrt{8640}}{2}\right)\left(x - \dfrac{-40 - \sqrt{8640}}{2}\right) \leq 0$

$[-6647.6, 2647.6]$

Production must be 2647 sweaters or fewer.

44. Revenue = $p(120 - 10p)$

$120p - 10p^2 > 350$

$10p^2 - 120p + 350 < 0$

$10(p - 5)(p - 7) < 0$

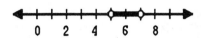

0   2   4   6   8

$(5, 7)$

The nursery should price rose food between \$5 and \$7 per box.

46. $x^2 + x + 20 \geq 0$

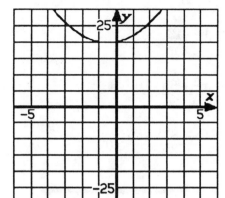

Solution: $(-\infty, +\infty)$
(all real numbers)

48. $x^2 - 2x < 3$ or $x^2 - 2x - 3 < 0$

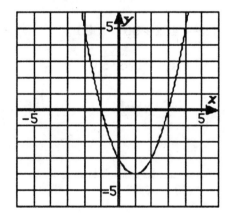

Solution: $(-1, 3)$

50. $x^2 - 6x + 9 > 0$

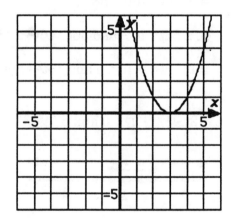

Solution: $(-\infty, 3) \cup (3, +\infty)$

52. $D = 4k^2 - 20$

$4k^2 - 20 > 0$

$4(k - \sqrt{5})(k + \sqrt{5}) > 0$

k must be less than $-\sqrt{5}$ or greater than $\sqrt{5}$.

54. $D = 16k^2 - 24k$

$16k^2 - 24k > 0$

$8k(2k - 3) > 0$

k must be greater than $\frac{3}{2}$ or less than 0.

56. $kx^2 - 2kx + (2k - 4) = 0$

$D = 4k^2 - 4k(2k - 4) = -4k^2 + 16k > 0$

$-4k(k - 4) > 0$

k must be greater than 0 and less than 4.

Exercise 10.4

2.

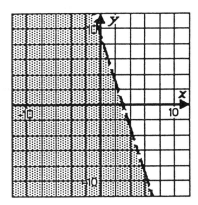

4.

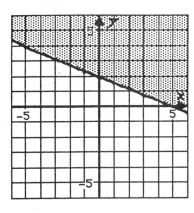

6.

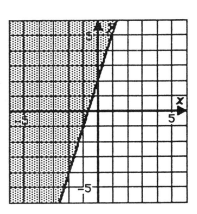

8.

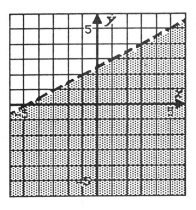

10.

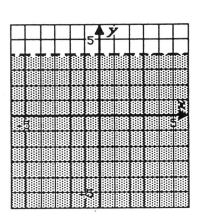

12.

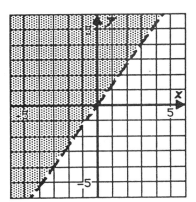

157

14.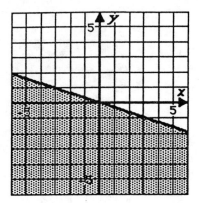

16.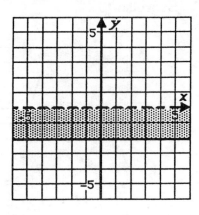

18.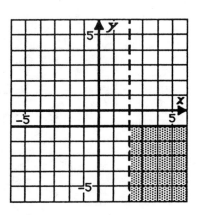

20.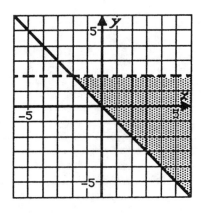

22.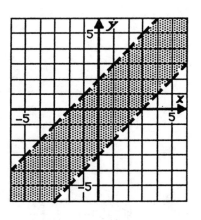

24.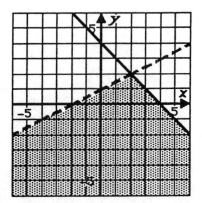

26.

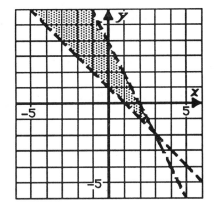

28.

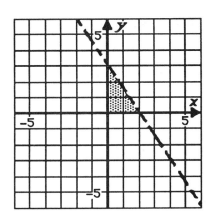

Solving the systems

x = 0	x = 0	y = 0
y = 0	3x + 2y = 6	3x + 2y = 6

yields the vertices (0,0), (0,3), (2,0)

30.

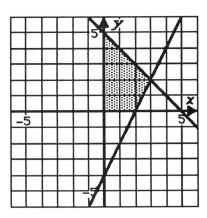

Solving the systems

x = 0	x = 0	y = 0
y = 0	x + y = 5	y - 2x = -4

x + y = 5
y - 2x = -4

yields the vertices (0,0), (0,5), (2,0), (3,2)

32.

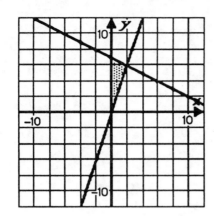

Solving the systems

$x = 0$	$x = 0$	$y = 3x$
$y = 3x$	$2y + x = 14$	$2y + x = 14$

yields the vertices $(0,0)$, $(0,7)$, $(2,6)$

34.

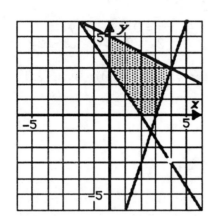

Solving the systems

$x = 0$	$x = 0$	$y = 0$
$2y + 3x = 6$	$2y + x = 10$	$2y + 3x = 6$

$y = 0$	$2y + x = 10$
$y = 3x - 9$	$y = 3x - 9$

yields the vertices $(0,3)$, $(0,5)$, $(2,0)$, $(3,0)$, $(4,3)$

36.

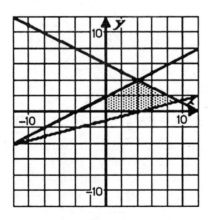

Solving the systems

$x = 0$	$x = 0$	$2y + x = 12$
$y = 0$	$4y = 2x + 8$	$4y = 2x + 8$

$y = 0$	$2y + x = 12$
$x = 4y + 4$	$x = 4y + 4$

yields the vertices $(0,0)$, $(0,2)$, $(4,4)$, $(4,0)$, $\left(\dfrac{28}{3}, \dfrac{4}{3}\right)$

CHAPTER 11

Exercise 11.1

2. $d = \sqrt{(5 - (-1))^2 + (9 - 1)^2} = \sqrt{36 + 64} = \sqrt{100} = 10$

$\overline{x} = \dfrac{-1 + 5}{2}$, $\overline{y} = \dfrac{1 + 9}{2}$, midpoint = $(2, 5)$

4. $d = \sqrt{(-1 - 5)^2 + (1 - (-4))^2} = \sqrt{36 + 25} = \sqrt{61}$

$\overline{x} = \dfrac{5 + (-1)}{2}$, $\overline{y} = \dfrac{-4 + 1}{2}$, midpoint = $\left(2, \dfrac{-3}{2}\right)$

6. $d = \sqrt{(-2 - (-2))^2 + (3 - (-5))^2} = \sqrt{0 + 64} = 8$

$\overline{x} = \dfrac{-2 + (-2)}{2}$, $\overline{y} = \dfrac{-5 + 3}{2}$, midpoint = $(-2, -1)$

8.

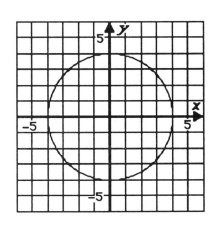

10.

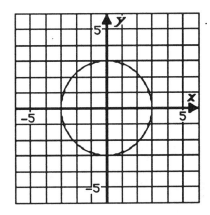

12.

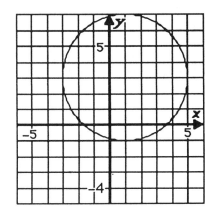

14.

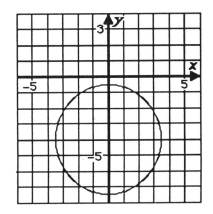

161

16.

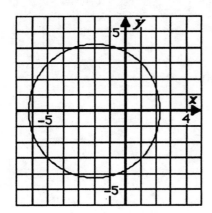

18.

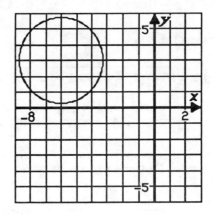

20. $x^2 + y^2 - 6x + 2y = 4$

$x^2 - 6x + 9 + y^2 + 2y + 1 = 4 + 9 + 1$

$(x - 3)^2 + (y + 1)^2 = 14$

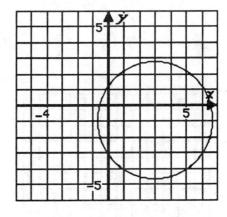

22. $x^2 + y^2 - 10y = 2$

$x^2 + y^2 - 10y + 25 = 2 + 25$

$(x - 0)^2 + (y - 5)^2 = 27$

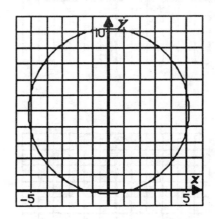

162

24. $x^2 + y^2 - 6x = 0$

$x^2 - 6x + 9 + y^2 = 0 + 9$

$(x - 3)^2 + (y - 0)^2 = 9$

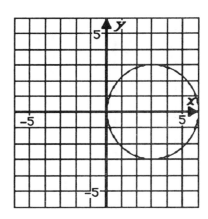

26. $(x - 4)^2 + (y - (-3))^2 = (2\sqrt{6})^2$ or $(x - 4)^2 + (y + 3)^2 = 24$

28. Tangent to the y-axis and center at $(1,7)$ implies $r = 1$: $(x - 1)^2 + (y - 7)^2 = 1$

30. $\overline{x} = \dfrac{3 + (-5)}{2}$, $\overline{y} = \dfrac{6 + 2}{2}$, center is at $(-1, 4)$; $d = \sqrt{(-1 - 3)^2 + (4 - 6)^2} = \sqrt{20}$.

$(x - (-1))^2 + (y - 4)^2 = (\sqrt{20})^2$ or $(x + 1)^2 + (y - 4)^2 = 20$

32. $\overline{x} = \dfrac{1 + (-4)}{2}$, $\overline{y} = \dfrac{1 + (-2)}{2}$, center is at $\left(\dfrac{-3}{2}, \dfrac{-1}{2}\right)$; $d = \sqrt{\left(\dfrac{-3}{2} - 1\right)^2 + \left(\dfrac{-1}{2} - 1\right)^2} =$

$\sqrt{\dfrac{25}{4} + \dfrac{9}{4}} = \dfrac{\sqrt{34}}{2}$. $\left(x - \left(\dfrac{-3}{2}\right)\right)^2 + \left(y - \left(\dfrac{-1}{2}\right)\right)^2 = \left(\dfrac{\sqrt{34}}{2}\right)^2$ or $\left(x + \dfrac{3}{2}\right)^2 + \left(y + \dfrac{1}{2}\right)^2 = \dfrac{34}{4} = \dfrac{17}{2}$

34.

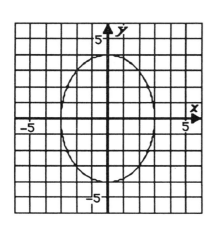

36.

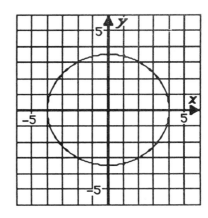

38.

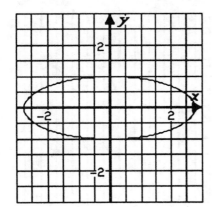

40.

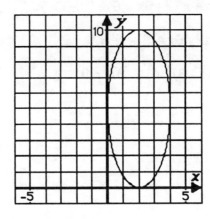

42.

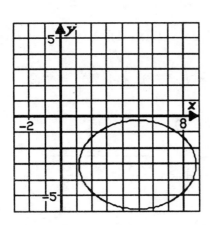

44.

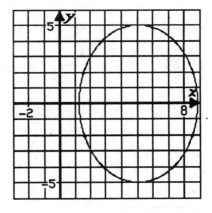

46. $x^2 + 16y^2 + 6x = 7$

$x^2 + 6x + 9 + 16y^2 = 7 + 9$

$(x + 3)^2 + 16y^2 = 16$

$\dfrac{(x + 3)^2}{16} + \dfrac{y^2}{1} = 1$

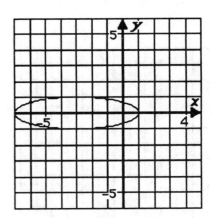

164

48. $16x^2 + 9y^2 + 64x - 18y = 71$

        $16(x^2 + 4x + 4) + 9(y^2 - 2y + 1) =$

        $\qquad\qquad\qquad 71 + 16(4) + 9(1)$

        $16(x + 2)^2 + 9(y - 1)^2 = 144$

        $\dfrac{(x + 2)^2}{9} + \dfrac{(y - 1)^2}{16} = 1$

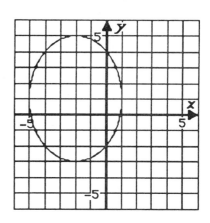

50. $2x^2 + y^2 - 16x + 6y = -11$

        $2(x^2 - 8x + 16) + (y^2 + 6y + 9) =$

        $\qquad\qquad\qquad -11 + 2(16) + 9$

        $2(x - 4)^2 + (y + 3)^2 = 30$

        $\dfrac{(x - 4)^2}{15} + \dfrac{(y + 3)^2}{30} = 1$

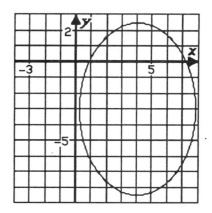

52. $5x^2 + 8y^2 - 20x + 16y = 12$

        $5(x^2 - 4x + 4) + 8(y^2 + 2y + 1) =$

        $\qquad\qquad\qquad 12 + 5(4) + 8(1)$

        $5(x - 2)^2 + 8(y + 1)^2 = 40$

        $\dfrac{(x - 2)^2}{8} + \dfrac{(y + 1)^2}{5} = 1$

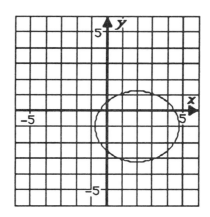

54. $x^2 + 10y^2 + 4x + 20y = -4$

        $(x^2 + 4x + 4) + 10(y^2 + 2y + 1) =$

        $\qquad\qquad\qquad -4 + 4 + 10(1)$

        $(x + 2)^2 + 10(y + 1)^2 = 10$

        $\dfrac{(x + 2)^2}{10} + \dfrac{(y + 1)^2}{1} = 1$

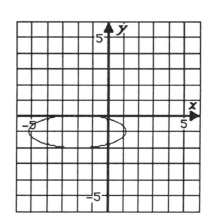

56. $\dfrac{(x-2)^2}{9} + \dfrac{(y-3)^2}{16} = 1$

58. $\bar{x} = \dfrac{3+3}{2}$, $\bar{y} = \dfrac{7+(-1)}{2}$, so the center is at $(3,3)$ and $b = 4$. The major axis has length 10, so $a = 5$. The covertices determine the minor axis, which is vertical. The equation is $\dfrac{(x-3)^2}{25} + \dfrac{(y-3)^2}{16} = 1$.

60. $\bar{x} = \dfrac{-3+9}{2}$, $\bar{y} = \dfrac{-5+(-5)}{2}$, so the center is at $(3, -5)$. The vertices are 6 units from the center, and the covertices are 5 units from the center, so $a = 6$ and $b = 5$. The vertices determine the major axis, which is horizontal. The equation is $\dfrac{(x-3)^2}{36} + \dfrac{(y+5)^2}{25} = 1$.

62. $s_1 = \sqrt{(8-(-1))^2 + (-7-5)^2} = \sqrt{81+144} = 15$; $s_2 = \sqrt{(4-8)^2 + (1-(-7))^2} = \sqrt{16+64} = \sqrt{80}$; $s_3 = \sqrt{(-1-4)^2 + (5-1)^2} = \sqrt{25+16} = \sqrt{41}$. The perimeter $= 15 + 4\sqrt{5} + \sqrt{41}$

64. $s_1 = \sqrt{(6-0)^2 + (0-0)^2} = \sqrt{36} = 6$; $s_2 = \sqrt{(3-6)^2 + (3-0)^2} = \sqrt{9+9} = \sqrt{18}$; $s_3 = \sqrt{(0-3)^2 + (0-3)^2} = \sqrt{9+9} = \sqrt{18}$. Since $s_2 = s_3$, the triangle is isosceles, and since $s_2^2 + s_3^2 = (\sqrt{18})^2 + (\sqrt{18})^2 = 18 + 18 = 36 = s_1^2$, the triangle is right.

66. $c = 0$ Thus $c = 0$, $36 + 6a = 0$ or $a = -6$, and $64 + 8b = 0$ or $b = -8$. The
$36 + 6a + c = 0$
$64 + 8b + c = 0$ required equation is $x^2 + y^2 - 6x - 8y = 0$.

68. Since A, E, and C all lie on the same horizontal line, and B, D, and C all lie on the same vertical line, A, E, and C must all have y-coordinate y_1, and B, D, and C must all have x-coordinate x_2, so $E = (\bar{x}, y_1)$, $D = (x_2, \bar{y})$, and $C = (x_2, y_1)$. We want $\bar{x}$ to satisfy the property that $AC = 2AE$, or $\sqrt{(x_1 - x_2)^2 + (y_1 - y_1)^2} = 2\sqrt{(x_1 - \bar{x})^2 + (y_1 - y_1)^2}$, which implies $x_1 - x_2 = 2(x_1 - \bar{x})$, $2\bar{x} = x_1 + x_2$, and $\bar{x} = \dfrac{x_1 + x_2}{2}$. A similar argument starting with $BC = 2BD$ and $\sqrt{(x_2 - x_2)^2 + (y_2 - y_1)^2} = 2\sqrt{(x_2 - x_2)^2 + (y_2 - \bar{y})^2}$ shows that $\bar{y} = \dfrac{y_1 + y_2}{2}$.

Exercise 11.2

2. $4p = 4$, so $p = 1$;

focus: $(1,0)$; directrix: $x = -1$

4. $-4p = -18$, so $p = \dfrac{9}{2}$;

focus: $\left(0, \dfrac{-9}{2}\right)$; directrix: $y = \dfrac{9}{2}$

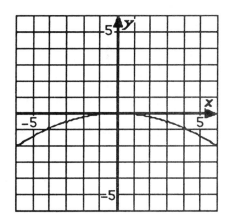

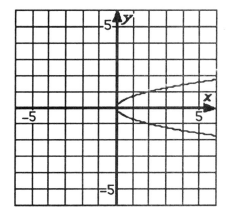

6. $x^2 = \dfrac{3}{4} y$, so $4p = \dfrac{3}{4}$ and $p = \dfrac{3}{16}$;

focus: $\left(0, \dfrac{3}{16}\right)$; directrix: $y = \dfrac{-3}{16}$

8. $y^2 = \dfrac{1}{2} x$, so $4p = \dfrac{1}{2}$ and $p = \dfrac{1}{8}$;

focus: $\left(\dfrac{1}{8}, 0\right)$; directrix: $x = \dfrac{-1}{8}$

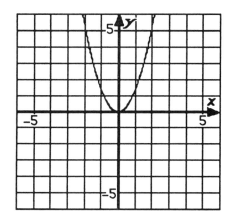

10. $x^2 = \dfrac{-5}{3} y$, so $-4p = \dfrac{-5}{3}$ and $p = \dfrac{5}{12}$;

focus: $\left(0, \dfrac{-5}{12}\right)$; directrix: $y = \dfrac{5}{12}$

12.

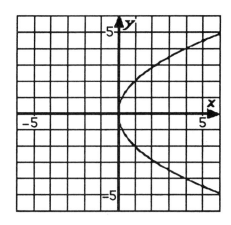

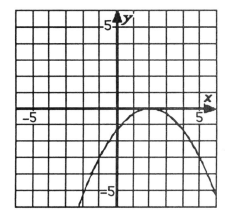

14.

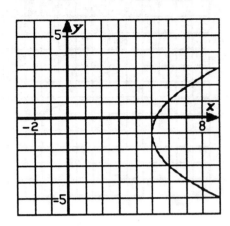

16.

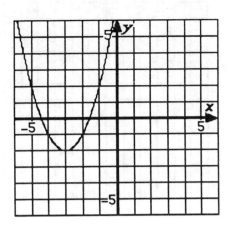

18. $x^2 - 8x - y + 2 = 0$

$x^2 - 8x + 16 = y - 2 + 16$

$(x - 4)^2 = 1(y + 14)$

20. $3y^2 + 6y - x + 4 = 0$

$3(y^2 + 2y + 1) = x - 4 + 3(1)$

$3(y + 1)^2 = (x - 1)$

$(y + 1)^2 = \frac{1}{3}(x - 1)$

22. $2x^2 - 5x + y + 1 = 0$

$2\left(x^2 - \frac{5}{2}x + \frac{25}{16}\right) = -y - 1 + 2\left(\frac{25}{16}\right) = -y - 1 + \frac{25}{8}$

$2\left(x - \frac{5}{4}\right)^2 = -1\left(y - \frac{17}{8}\right)$

$\left(x - \frac{5}{4}\right)^2 = \frac{-1}{2}\left(y - \frac{17}{8}\right)$

24. $y^2 - 4y + 8x + 6 = 0$

$y^2 - 4y + 4 = -8x - 6 + 4$

$(y - 2)^2 = -8\left(x + \frac{1}{4}\right)$

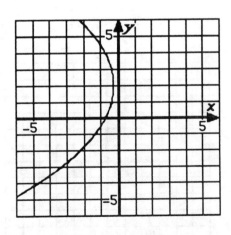

26. $4x^2 + 4x = 8y - 5$

$$4\left(x^2 + x + \frac{1}{4}\right) = 8y - 5 + 4\left(\frac{1}{4}\right)$$

$$4\left(x + \frac{1}{2}\right)^2 = 8\left(y - \frac{1}{2}\right)$$

$$\left(x + \frac{1}{2}\right)^2 = 2\left(y - \frac{1}{2}\right)$$

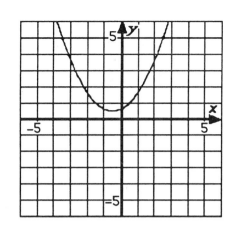

28. $9y^2 + 12y - 12x = 0$

$$9\left(y^2 + \frac{4}{3}y + \frac{4}{9}\right) = 12x + 4$$

$$9\left(y + \frac{2}{3}\right)^2 = 12\left(x + \frac{1}{3}\right)$$

$$\left(y + \frac{2}{3}\right)^2 = \frac{4}{3}\left(x + \frac{1}{3}\right)$$

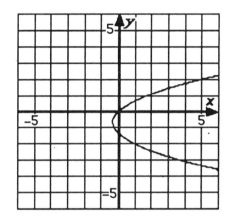

30. $p = 2$, the axis is vertical, so the equation is $x^2 = -8y$

32. $p = 4$, the axis is horizontal, so the equation is $y^2 = 16x$

34. $p = 2$, the axis is vertical, so the equation is $(x + 1)^2 = 8(y - 3)$

36. $p = 3$, the axis is horizontal, so the equation is $(y + 3)^2 = -12(x - 6)$

38. $p = 2$, the axis is vertical, so the equation is $(x + 4)^2 = -8(y - 1)$

40. The vertex is halfway between the focus and the directrix, so the vertex is

(-4, -2) and $p = 2$. The axis is horizontal, so the equation is $(y + 2)^2 = 8(x + 4)$.

42. The equation has the form $y^2 = 4px$, so $9 = 4p$, $p = \frac{9}{4}$, and the equation is $y^2 = 9x$.

44. The equation has the form $(x - 1)^2 = 4p(y + 5)$, so $(3 - 1)^2 = 4p(3 + 5)$, or $4 = 32p$.

Thus $p = \frac{1}{8}$ and the equation is $(x - 1)^2 = \frac{1}{2}(y + 5)$.

169

Exercise 11.3

2.

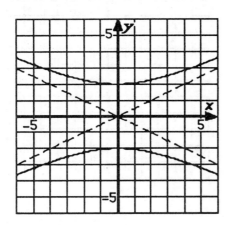

4.

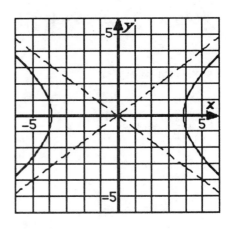

6.

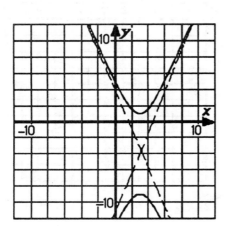

8.

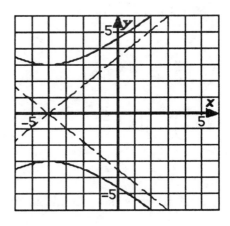

10.

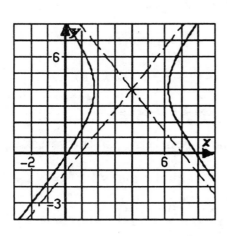

12.

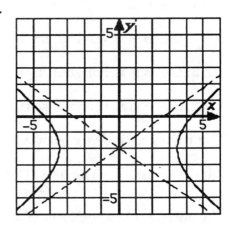

170

14. $9y^2 - 4x^2 - 72y - 24x = -72$

$9(y^2 - 8y + 16) - 4(x^2 + 6x + 9) =$

$$-72 + 9(16) - 4(9)$$

$9(y - 4)^2 - 4(x + 3)^2 = 36$

$$\frac{(y - 4)^2}{4} - \frac{(x + 3)^2}{9} = 1$$

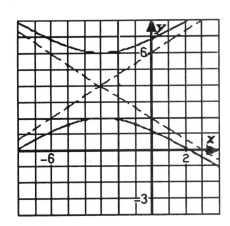

16. $16x^2 - 9y^2 + 54y = 225$

$16x^2 - 9(y^2 - 6y + 9) = 225 - 9(9)$

$16x^2 - 9(y - 3)^2 = 144$

$$\frac{x^2}{9} - \frac{(y - 3)^2}{16} = 1$$

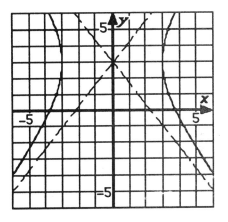

18. $9y^2 - 8x^2 + 72y + 16x = -64$

$9(y^2 + 8y + 16) - 8(x^2 - 2x + 1) =$

$$-64 + 9(16) - 8(1)$$

$9(y + 4)^2 - 8(x - 1)^2 = 72$

$$\frac{(y + 4)^2}{8} - \frac{(x - 1)^2}{9} = 1$$

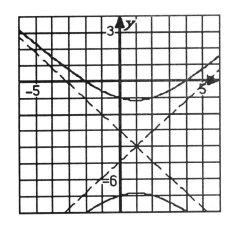

171

20. $10y^2 - 5x^2 + 30x = 95$

$10y^2 - 5(x^2 - 6x + 9) = 95 - 5(9)$

$10y^2 - 5(x - 3)^2 = 50$

$\dfrac{y^2}{5} - \dfrac{(x - 3)^2}{10} = 1$

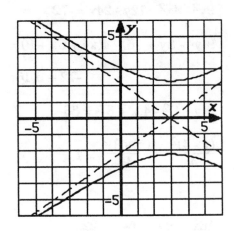

22. $y^2 = 6 - 4x^2$

$4x^2 + y^2 = 6$

$\dfrac{x^2}{1.5} + \dfrac{y^2}{6} = 1$

an ellipse with center at the origin with the major axis of length $2\sqrt{6}$ parallel to the y-axis with vertices $(0, \pm\sqrt{6})$ and covertices $(\pm\sqrt{1.5}, 0)$

24. $x^2 + 2y - 4 = 0$

$x^2 = -2(y - 2)$

a parabola opening downward from the vertex $(0, 2)$

26. $6x^2 = 8 - 6y^2$

$6x^2 + 6y^2 = 8$

$x^2 + y^2 = \dfrac{4}{3}$

a circle with center at the origin and radius $\dfrac{2}{\sqrt{3}}$

28. $2x^2 = 5 + 4y^2$

$2x^2 - 4y^2 = 5$

$\dfrac{x^2}{2.5} - \dfrac{y^2}{1.25} = 1$

a hyperbola with center at the origin and vertices $(\pm\sqrt{2.5}, 0)$

30. $y^2 = 6 - \dfrac{2x^2}{3}$

$\dfrac{2x^2}{3} + y^2 = 6$

$\dfrac{x^2}{9} + \dfrac{y^2}{6} = 1$

an ellipse with center at the origin, major axis of length 6 parallel to the x-axis with vertices $(\pm 3, 0)$ and covertices $(0, \pm\sqrt{6})$

32. $\dfrac{x^2}{4} = 4 + 6y^2$

$\dfrac{x^2}{4} - 6y^2 = 4$

$\dfrac{x^2}{16} - \dfrac{y^2}{2/3} = 1$

a hyperbola with center at the origin and vertices $(\pm 4, 0)$

34. $\dfrac{(y-2)^2}{4} - \dfrac{(x+3)^2}{8} = 1$

a hyperbola with center at (-3, 2) and vertices (-3, 0) and (-3, 4)

36. $\dfrac{(x+3)^2}{4} + \dfrac{y^2}{12} = 1$

an ellipse with center at (-3, 0) with major axis of length $4\sqrt{3}$ parallel to the y-axis with vertices (-3, $\pm 2\sqrt{3}$) and covertices (-5, 0) and (-1, 0)

38. $y^2 - 4x^2 + 2y - x = 0$

$(y^2 + 2y + 1) - 4\left(x^2 + \dfrac{1}{4}x + \dfrac{1}{64}\right) = 0 + 1 - 4\left(\dfrac{1}{64}\right)$

$(y+1)^2 - 4\left(x + \dfrac{1}{8}\right)^2 = \dfrac{15}{16}$

$\dfrac{(y+1)^2}{\frac{15}{16}} - \dfrac{\left(x + \frac{1}{8}\right)^2}{\frac{15}{64}} = 1$

a hyperbola with center at $\left(\dfrac{-1}{8}, -1\right)$

and vertices $\left(\dfrac{-1}{8}, -1 \pm \dfrac{\sqrt{15}}{4}\right)$

40. $y - 2 = \dfrac{(x+4)^2}{4}$

$4(y - 2) = (x + 4)^2$

a parabola opening upward from the vertex (-4, 2)

42. $4x^2 - y^2 = 0$

$(2x - y)(2x + y) = 0$

$2x - y = 0 \text{ or } 2x + y = 0$

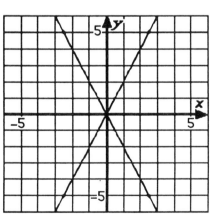

44.

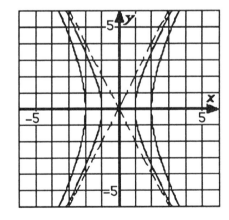

46. $\dfrac{(x-6)^2}{1} - \dfrac{(y+2)^2}{16} = 1$

48. $\dfrac{(x+5)^2}{36} - \dfrac{(y+2)^2}{4} = 1$ OR

$\dfrac{(y+4)^2}{4} - \dfrac{(x-1)^2}{36} = 1$

2. $y = x^2 - 2x + 1$
    $y = 3 - x$

    $3 - x = x^2 - 2x + 1$

    $x^2 - x - 2 = 0$

    $(x - 2)(x + 1) = 0$

    $x = 2$ or $x = -1$

    the solutions are $(2,1)$ and $(-1,4)$

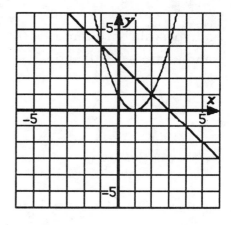

4. $x^2 + 2y^2 = 12$
    $2x - 2 = y$

    $x^2 + 2(2x - 2)^2 = 12$

    $9x^2 - 16x - 4 = 0$

    $(9x + 2)(x - 2) = 0$

    $x = \dfrac{-2}{9}$ or $x = 2$

    the solutions are $\left(\dfrac{-2}{9}, \dfrac{-22}{9}\right)$ and $(2,2)$

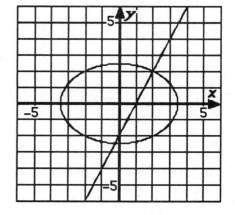

6. $2x - 9 = y$
    $xy = -4$

    $x(2x - 9) = -4$

    $2x^2 - 9x + 4 = 0$

    $(2x - 1)(x - 4) = 0$

    $x = \dfrac{1}{2}$ or $x = 4$

    the solutions are $\left(\dfrac{1}{2}, -8\right)$ and $(4,-1)$

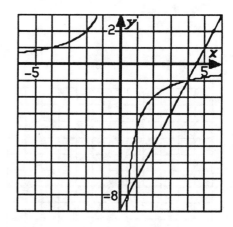

8. $x^2 - y^2 = 35$

    $y = \dfrac{6}{x}$

    $x^2 - \dfrac{36}{x^2} = 35$

    $x^4 - 35x^2 - 36 = 0$

    $(x^2 - 36)(x^2 + 1) = 0$

    $(x - 6)(x + 6)(x^2 + 1) = 0$

    $x^2 + 1$ is never 0, so $x = 6$ or $x = -6$

    the solutions are (6,1) and (-6,-1)

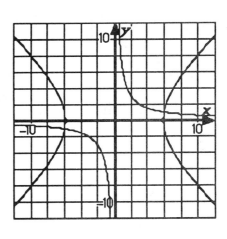

10. $2x^2 - 4y^2 = 12$
    $x = 4$

    $32 - 4y^2 = 12$

    $4y^2 = 20$

    $y^2 = 5$

    $y = \pm \sqrt{5}$

    the solutions are $(4, \sqrt{5})$ and $(4, -\sqrt{5})$

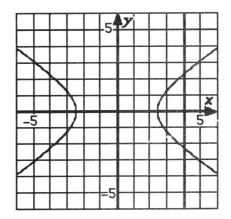

12. $x^2 + 9y^2 = 36$
    $x = 2y - 8$

    $(2y - 8)^2 + 9y^2 = 36$

    $13y^2 - 32y + 28 = 0$

    D = 1024 - 1456 = -432, so this
    system has no solutions

14. $x^2 - 2x + y^2 = 3$
    $y = 4 - 2x$

    $x^2 - 2x + (4 - 2x)^2 = 3$

    $5x^2 - 18x + 13 = 0$

    $(5x - 13)(x - 1) = 0$

    $x = \dfrac{13}{5}$ or $x = 1$

    the solutions are $\left(\dfrac{13}{5}, \dfrac{-6}{5}\right)$ and (1,2)

16. $2x^2 + xy + y^2 = 9$
    $x = 3y - 9$

    $2(3y - 9)^2 + (3y - 9)y + y^2 = 9$

18. $x^2 + 4y^2 = 52$
    $\dfrac{-x^2 - y^2 = -25}{3y^2 = 27}$

    $y^2 = 9$, so $y = \pm 3$

175

$(y - 3)(22y - 51) = 0$

$y = 3$ or $y = \dfrac{51}{22}$

the solutions are $(0,3)$ and $\left(\dfrac{-45}{22}, \dfrac{51}{22}\right)$

the solutions are $(4,3)$, $(4,-3)$, $(-4,3)$, and $(-4,-3)$

20.
$$9x^2 + 16y^2 = 100$$
$$\underline{-9x^2 - 9y^2 = -72}$$
$$7y^2 = 28$$

$y^2 = 4$, so $y = \pm 2$

$x^2 + 4 = 8$, $x^2 = 4$, $x = \pm 2$

the solutions are $(2,2)$, $(2,-2)$, $(-2,2)$, $(-2,-2)$

22.
$$x^2 + 4y^2 = 25$$
$$\underline{-16x^2 - 4y^2 = -100}$$
$$-15x^2 = -75$$

$x^2 = 5$, so $x = \pm\sqrt{5}$

$5 + 4y^2 = 25$, $y^2 = 5$, $y = \pm\sqrt{5}$

the solutions are $(\sqrt{5}, \sqrt{5})$, $(\sqrt{5}, -\sqrt{5})$, $(-\sqrt{5}, \sqrt{5})$, $(-\sqrt{5}, -\sqrt{5})$

24.
$$4x^2 + 3y^2 = 12$$
$$\underline{-x^2 - 3y^2 = -12}$$
$$3x^2 = 0$$

$x^2 = 0$, $x = 0$

$3y^2 = 12$, $y^2 = 4$, $y = \pm 2$

the solutions are $(0,2)$ and $(0,-2)$

26.
$$64y^2 + 20x^2 - 104 = 0$$
$$\underline{125y^2 - 20x^2 - 85 = 0}$$
$$189y^2 - 189 = 0$$

$y^2 = 1$, $y = \pm 1$

$16 + 5x^2 - 26 = 0$, $x^2 = 2$, $x = \pm\sqrt{2}$

the solutions are $(\sqrt{2}, 1)$, $(\sqrt{2}, -1)$, $(-\sqrt{2}, 1)$, $(-\sqrt{2}, -1)$

28.
$$x^2 + 2xy - y^2 = 14$$
$$\underline{-x^2 + y^2 = -8}$$
$$2xy = 6 \text{ or } xy = 3$$

$y = \dfrac{3}{x}$

$x^2 - \dfrac{9}{x^2} = 8$

$x^4 - 8x^2 - 9 = 0$

$(x^2 - 9)(x^2 + 1) = 0$

$(x - 3)(x + 3)(x^2 + 1) = 0$

$x^2 + 1$ is never 0, so $x = 3$ or $x = -3$

the solutions are $(3,1)$ and $(-3,-1)$

30.
$$2x^2 + xy - 2y^2 = 16$$
$$\underline{-2x^2 - 4xy + 2y^2 = -34}$$
$$-3xy = -18 \text{ or } xy = 6$$

$y = \dfrac{6}{x}$

$2x^2 + x\left(\dfrac{6}{x}\right) - 2\left(\dfrac{36}{x^2}\right) = 16$

$2x^4 - 10x^2 - 72 = 0$

$2(x^2 - 9)(x^2 + 4) = 0$

$2(x - 3)(x + 3)(x^2 + 4) = 0$

$x^2 + 4$ is never 0, so $x = 3$ or $x = -3$

the solutions are $(3,2)$ and $(-3,-2)$

32.

$$3x^2 - 2xy + 3y^2 = 34$$
$$\underline{-3x^2 \qquad\quad - 3y^2 = -51}$$
$$-2xy \qquad\quad = -17 \text{ or } y = \frac{17}{2x}$$

$$x^2 + \frac{289}{4x^2} = 17$$

$$4x^4 - 68x^2 + 289 = 0$$

$$(2x^2 - 17)^2 = 0$$

$$x = \pm \sqrt{\frac{17}{2}}$$

the solutions are $\left(\sqrt{\frac{17}{2}}, \sqrt{\frac{17}{2}}\right)$ and

$$\left(-\sqrt{\frac{17}{2}}, -\sqrt{\frac{17}{2}}\right)$$

34.

$$x^2 + 2xy - 8y^2 = 0 \text{ means}$$
$$(x + 4y)(x - 2y) = 0, \text{ so}$$

$$x = -4y \text{ or } x = 2y.$$

$$(-4y)^2 - (-4y)y + y^2 = 21 \text{ or}$$

$$(2y)^2 - (2y)y + y^2 = 21$$

$$21y^2 = 21 \text{ or } 3y^2 = 21$$

$$y^2 = 1 \text{ or } y^2 = 7; \; y = \pm 1 \text{ or } y = \pm\sqrt{7}$$

the solutions are $(-4,1)$, $(4,-1)$,

$$(2\sqrt{7}, \sqrt{7}), (-2\sqrt{7}, -\sqrt{7})$$

36.

$$-2x^2 - xy + y^2 = 0$$
$$\underline{6x^2 + xy - y^2 = 1}$$
$$4x^2 \qquad\qquad = 1$$

$$x^2 = \frac{1}{4}, \; x = \pm\frac{1}{2}$$

$$\frac{1}{2} + \frac{1}{2}y - y^2 = 0 \text{ or } \frac{1}{2} - \frac{1}{2}y - y^2 = 0$$

$$2y^2 - y - 1 = 0 \text{ or } 2y^2 + y - 1 = 0$$

$$(2y + 1)(y - 1) = 0 \text{ or } (2y - 1)(y + 1) = 0$$

$$y = \frac{-1}{2} \text{ or } y = 1 \text{ or } y = \frac{1}{2} \text{ or } -1$$

the solutions are $\left(\frac{1}{2}, 1\right)$, $\left(\frac{1}{2}, \frac{-1}{2}\right)$,

$$\left(\frac{-1}{2}, \frac{1}{2}\right), \left(\frac{-1}{2}, -1\right)$$

38.

the two numbers: x and y

$$x + y = 6$$
$$xy = \frac{35}{4}$$
$$x(6 - x) = \frac{35}{4}$$

$$24x - 4x^2 = 35$$

$$4x^2 - 24x + 35 = 0$$

$$(2x - 5)(2x - 7) = 0$$

$$x = \frac{5}{2} \text{ or } x = \frac{7}{2}$$

the numbers are $\frac{5}{2}$ and $\frac{7}{2}$

40.

length: L
width: W

$$LW = 216$$
$$2L + 2W = 60$$

42.

$$PV = 30$$
$$(P + 4)(V - 2) = 30$$

$$PV + 4V - 2P - 8 = 30$$

$$V = \frac{30}{P}$$

$$L = 30 - W$$

$$(30 - W)W = 216$$

$$W^2 - 30W + 216 = 0$$

$$(W - 18)(W - 12) = 0$$

$$W = 18 \text{ or } 12$$

the rectangle is 18 x 12 ft.

$$P\left(\frac{30}{P}\right) + 4\left(\frac{30}{P}\right) - 2P - 38 = 0$$

$$30 + \frac{120}{P} - 2P - 38 = 0$$

$$2P^2 + 8P - 120 = 0$$

$$2(P + 10)(P - 6) = 0$$

Pressure cannot be negative, so the pressure is 6 lb. per sq. in. and the volume is 5 cu. in.

44.   Since all possible solutions must satisfy both equations, the apparent solutions (2, -2) and (-2, 2) are extraneous solutions, since they satisfy equation (1) but not equation (2). The only solutions are the ordered pairs (2, 2) and (-2, -2), as is clear from the graph at the right.

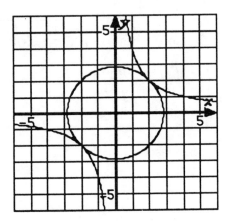

CHAPTER 12

Exercise 12.1

2. $s_1 = 2(1) - 3 = -1$

$s_2 = 2(2) - 3 = 1$

$s_3 = 2(3) - 3 = 3$

$s_4 = 2(4) - 3 = 5$

The first four terms are -1, 1, 3 , and 5

4. $s_1 = \dfrac{3}{1^2 + 1} = \dfrac{3}{2}$

$s_2 = \dfrac{3}{2^2 + 1} = \dfrac{3}{5}$

$s_3 = \dfrac{3}{3^2 + 1} = \dfrac{3}{10}$

$s_4 = \dfrac{3}{4^2 + 1} = \dfrac{3}{17}$

The first four terms are $\dfrac{3}{2}, \dfrac{3}{5}, \dfrac{3}{10}$, and $\dfrac{3}{17}$

6. $s_1 = \dfrac{1}{2(1) - 1} = 1$

$s_2 = \dfrac{2}{2(2) - 1} = \dfrac{2}{3}$

$s_3 = \dfrac{3}{2(3) - 1} = \dfrac{3}{5}$

$s_4 = \dfrac{4}{2(4) - 1} = \dfrac{4}{7}$

The first four terms are $1, \dfrac{2}{3}, \dfrac{3}{5}$, and $\dfrac{4}{7}$

8. $s_1 = \dfrac{5}{1(1 + 1)} = \dfrac{5}{2}$

$s_2 = \dfrac{5}{2(2 + 1)} = \dfrac{5}{6}$

$s_3 = \dfrac{5}{3(3 + 1)} = \dfrac{5}{12}$

$s_4 = \dfrac{5}{4(4 + 1)} = \dfrac{5}{20} = \dfrac{1}{4}$

The first four terms are $\dfrac{5}{2}, \dfrac{5}{6}, \dfrac{5}{12}$, and $\dfrac{1}{4}$

10. $s_1 = (-1)^{1+1} = 1$

$s_2 = (-1)^{2+1} = -1$

$s_3 = (-1)^{3+1} = 1$

$s_4 = (-1)^{4+1} = -1$

The first four terms are 1, -1, 1, and -1

12. $s_1 = (-1)^{1-1}3^{1+1} = 9$

$s_2 = (-1)^{2-1}3^{2+1} = -27$

$s_3 = (-1)^{3-1}3^{3+1} = 81$

$s_4 = (-1)^{4-1}3^{4+1} = -243$

The first four terms are 9, -27, 81, and -243

14. $s_{11} = \sqrt{12} = 2\sqrt{3}$

16. $s_9 = \log 10 = 1$

18. $s_{17} = \dfrac{17 + 1}{17 - 1} = \dfrac{18}{16} = \dfrac{9}{8}$

20. $s_2 = 6 - 4 = 2, s_3 = 2 - 4 = -2, s_4 = -2 - 4 = -6, s_5 = -6 - 4 = -10$.

The first five terms are 6, 2, -2, -6, and -10.

22. $s_2 = \dfrac{2}{3}(27) = 18, s_3 = \dfrac{2}{3}(18) = 12, s_4 = \dfrac{2}{3}(12) = 8, s_5 = \dfrac{2}{3}(8) = \dfrac{16}{3}$.

The first five terms are 27, 18, 12, 8, and $\dfrac{16}{3}$.

24. $s_2 = \left(\dfrac{1 + 1}{1}\right)1 = 2, s_3 = \left(\dfrac{2 + 1}{2}\right)2 = 3, s_4 = \left(\dfrac{3 + 1}{3}\right)3 = 4, s_5 = \left(\dfrac{4 + 1}{4}\right)4 = 5$.

The first five terms are 1, 2, 3, 4, and 5.

26. $s_1 = \$21,000, s_{n+1} = 1.05s_n.$ $s_2 = 1.05(21,000) = 22,050, s_3 = 1.05(22,050) = 23,152.50,$

$s_4 = 1.05(23,152.50) = 24,310.13, s_5 = 1.05(24,310.13) = 25,525.63.$

The first five terms are $21,000, $22,050, $23,152.50, $24,310.13, and $25,525.63.

28. $s_1 = \$18{,}000$, $s_{n+1} = s_n + 0.005s_n - 400 = 1.005s_n - 400$.

$s_2 = 1.005(18{,}000) - 400 = 17{,}690$, $s_3 = 1.005(17{,}690) - 400 = 17{,}378.45$, $s_4 =$

$1.005(17{,}378.45) - 400 = 17{,}065.34$, $s_5 = 1.005(17{,}065.34) - 400 = 16{,}750.67$.

The first five terms are $18,000, $17,690, $17,378.45, $17,065.34, and $16,750.67.

30. $s_1 = 1$, $s_2 = 2$, $s_3 = \dfrac{5}{3}$, $s_4 = \dfrac{7}{4}$, $s_5 = \dfrac{19}{11}$, $s_{10} = 1.7320574$, $s_{100} = 1.7320508$;

s_n approaches 1.7320508, which is $\sqrt{3}$.

32. $s_1 = 8$, $s_2 = \dfrac{3}{2}$, $s_3 = 0.7906$, $s_4 = 0.66906$, $s_5 = 0.64596$, $s_{10} = 0.64039$, $s_{20} =$

0.6403882, $s_{100} = 0.6403882$; s_n approaches 0.6403882.

34. $s_1 = 1$, $s_2 = 5$, $s_3 = 3.4$, $s_4 = 3.023529$, $s_5 = 3.0000916$, $s_{10} = 3.0000000$; s_n

approaches 3.

36. a. $s_1 = 2$, $s_2 = 1$, $s_3 = 3$, $s_4 = 4$, $s_5 = 7$, $s_6 = 11$, $s_7 = 18$, $s_8 = 29$, $s_9 = 47$, $s_{10} = 76$.

b. $s_2^2 - s_1 s_3 = -5$, $s_3^2 - s_2 s_4 = 5$, $s_4^2 - s_3 s_5 = -5$, $s_5^2 - s_4 s_6 = 5$, $s_6^2 - s_5 s_7 = -5$,

$s_7^2 - s_6 s_8 = 5$, $s_8^2 - s_7 s_9 = -5$, $s_9^2 - s_8 s_{10} = 5$. For n even, the quantity equals

5, and for n odd, it equals -5. In symbols, $s_{n+1}^2 - s_n s_{n+2} = (-1)^n(5)$.

Exercise 12.2

2. $S_8 = (-1) + 2 + (-3) + 4 + (-5) + 6 + (-7) + 8 = 4$

4. $S_6 = 1 + 2 + 4 + 8 + 16 + 32 = 63$ 6. $S_5 = 1 + \dfrac{1}{4} + \dfrac{1}{9} + \dfrac{1}{16} + \dfrac{1}{25} = \dfrac{5269}{3600} \approx 1.4636$

8. $\displaystyle\sum_{j=1}^{100} (j + 1)(j + 2)$ 10. $\displaystyle\sum_{k=2}^{9} k^3 x^{k+1}$

12. for $i = 1$, $3(1) - 2 = 1$ 14. for $j = 2$, $2^2 + 1 = 5$
for $i = 2$, $3(2) - 2 = 4$ for $j = 3$, $3^2 + 1 = 10$
for $i = 3$, $3(3) - 2 = 7$ for $j = 4$, $4^2 + 1 = 17$
                                       for $j = 5$, $5^2 + 1 = 26$
expanded form is $1 + 4 + 7$ for $j = 6$, $6^2 + 1 = 37$

                                       expanded form is $5 + 10 + 17 + 26 + 37$

16. for $i = 2$, $\frac{2}{2}(2 + 1) = 3$

for $i = 3$, $\frac{3}{2}(3 + 1) = 6$

for $i = 4$, $\frac{4}{2}(4 + 1) = 10$

for $i = 5$, $\frac{5}{2}(5 + 1) = 15$

for $i = 6$, $\frac{6}{2}(6 + 1) = 21$

expanded form is $3 + 6 + 10 + 15 + 21$

18. for $i = 3$, $\frac{(-1)^{3+1}}{3 - 2} = 1$

for $i = 4$, $\frac{(-1)^{4+1}}{4 - 2} = \frac{-1}{2}$

for $i = 5$, $\frac{(-1)^{5+1}}{5 - 2} = \frac{1}{3}$

expanded form is $1 + \left(\frac{-1}{2}\right) + \frac{1}{3}$

20. for $j = 1$, $\frac{1}{1} = 1$

for $j = 2$, $\frac{1}{2} = \frac{1}{2}$

for $j = 3$, $\frac{1}{3} = \frac{1}{3}$

for $j = 4$, $\frac{1}{4} = \frac{1}{4}$

expanded form is $1 + \frac{1}{2} + \frac{1}{3} + \frac{1}{4} + \ldots$

22. for $k = 0$, $\frac{0}{1 + 0} = 0$

for $k = 1$, $\frac{1}{1 + 1} = \frac{1}{2}$

for $k = 2$, $\frac{2}{1 + 2} = \frac{2}{3}$

for $k = 3$, $\frac{3}{1 + 3} = \frac{3}{4}$

expanded form is $0 + \frac{1}{2} + \frac{2}{3} + \frac{3}{4} + \frac{4}{5} + \ldots$

24. $\displaystyle\sum_{i=1}^{4} 2i$

26. $\displaystyle\sum_{k=1}^{5} x^{2k+1}$ or $\displaystyle\sum_{k=2}^{6} x^{2k-1}$

28. $\displaystyle\sum_{j=1}^{5} j^3$

30. $\displaystyle\sum_{i=1}^{\infty} \frac{i + 1}{i}$ or $\displaystyle\sum_{i=2}^{\infty} \frac{i}{i - 1}$

32. $\displaystyle\sum_{k=1}^{\infty} \frac{2k + 1}{2k - 1}$

34. $\displaystyle\sum_{i=1}^{\infty} \frac{3^{i-1}}{2i}$ or $\displaystyle\sum_{i=0}^{\infty} \frac{3^i}{2i + 2}$

36. a. d_k: distance fallen during kth second; $d_k = (2k - 1)16 = 32k - 16$

b. $S_n = \displaystyle\sum_{k=1}^{n} d_k$; $S_1 = 16$, $S_2 = 16 + 48 = 64$, $S_3 = 64 + 80 = 144$, $S_4 = 144 + 112 = 256$. Since $S_1 = 16(1)$, $S_2 = 16(4)$, $S_3 = 16(9)$, and $S_4 = 16(16)$, we may guess that $S_n = 16n^2$.

38. $S_1 = s_1 = \log \frac{1}{2}$, $S_2 = s_1 + s_2 = \log \frac{1}{2} + \log \frac{2}{3} = \log \frac{1}{2}\left(\frac{2}{3}\right) = \log \frac{1}{3}$, $S_3 = s_1 + s_2 + s_3 =$

$\log \frac{1}{3} + \log \frac{3}{4} = \log \frac{1}{3}\left(\frac{3}{4}\right) = \log \frac{1}{4}$; $S_n = \log \frac{1}{n + 1}$

40. $S_1 = 1, S_2 = 1 + \frac{1}{4} = \frac{5}{4}, S_3 = \frac{5}{4} + \frac{1}{9} = \frac{49}{36}, S_4 = \frac{205}{144}, S_5 = \frac{5269}{3600}, S_{10} = 1.5497677, S_{20} =$

1.5961632, $S_{50} = 1.6251327$, $S_{100} = 1.6349839$, $S_{200} = 1.6399465$, $S_{500} = 1.6429361$,

$S_{2000} = 1.6444342$. $\sqrt{6S_1} = \sqrt{6}$, $\sqrt{6S_5} = 2.9634$, $\sqrt{6S_{10}} = 3.0494$, $\sqrt{6S_{50}} = 3.1226$,

$\sqrt{6S_{100}} = 3.1321$, $\sqrt{6S_{500}} = 3.1397$, $\sqrt{6S_{2000}} = 3.1411$. $\sqrt{6S_n}$ is approaching π.

Exercise 12.3

2. $s_1 = 7, s_2 = 7 + 3 = 10, s_3 = 10 + 3 = 13, s_4 = 13 + 3 = 16$

4. $s_1 = 8, s_2 = 8 + (-5) = 3, s_3 = 3 + (-5) = -2, s_4 = -2 + (-5) = -7$

6. $s_1 = \frac{2}{3}, s_2 = \frac{2}{3} + \frac{1}{3} = 1, s_3 = 1 + \frac{1}{3} = \frac{4}{3}, s_4 = \frac{4}{3} + \frac{1}{3} = \frac{5}{3}$

8. $s_1 = 6, s_2 = 6 + (1 - x) = 7 - x, s_3 = 7 - x + (1 - x) = 8 - 2x, s_4 = 8 - 2x + (1 - x) = 9 - 3x$

10. Next three terms: 4, 9, 14; $a = -6, d = 5$, so $s_n = -6 + (n - 1)5 = 5n - 11$

12. Next three terms: -30, -40, -50; $a = -10, d = -10$, so $s_n = -10 + (n - 1)(-10) = -10n$

14. Next three terms: $h + 10, h + 15, h + 20$; $a = h, d = 5$, so $s_n = h + (n - 1)5 =$

    $5n + (h - 5)$

16. Next three terms: $y + 2b, y + 4b, y + 6b$; $a = y - 2b, d = 2b$, so $s_n =$

    $(y - 2b) + (n - 1)2b = 2bn + (y - 4b)$

18. Next three terms: $q - 6b, q - 10b, q - 14b$; $a = q + 2b, d = -4b$, so $s_n =$

    $q + 2b + (n - 1)(-4b) = -4bn + (q + 6b)$

20. Next three terms: $7b, 9b, 11b$; $a = 3b, d = 2b$, so $s_n = 3b + (n - 1)(2b) = b(2n + 1)$

22. $a = -3, d = -9$, so $s_n = -3 + (n - 1)(-9)$ and $s_{10} = -3 + 9(-9) = -84$

24. $a = -5, d = 3$, so $s_n = -5 + (n - 1)3$ and $s_{17} = -5 + 16(3) = 43$

26. $a = \frac{3}{4}, d = \frac{5}{4}$, so $s_n = \frac{3}{4} + (n - 1)\left(\frac{5}{4}\right)$ and $s_{10} = \frac{3}{4} + 9\left(\frac{5}{4}\right) = \frac{48}{4} = 12$

28. $-16 = a + 4d, -46 = a + 19d$, so $-16 = a + 4d$
    $\underline{46 = -a + 19d}$
    $30 = -15d$, so $d = -2, a = -8$, and $a_{12} = -8 + 11(-2) = -30$

182

30. $a = 7, d = -4$, so $s_n = 7 + (n - 1)(-4) = -4n + 11$. Then $-81 = -4n + 11$, so $-92 = -4n$ and

$n = 23$. Thus -81 is the 23rd term in the progression.

32. The first term is 3, the last is $5 - 2(10) = -15$, so $S_{10} = \frac{10}{2}(3 + (-15)) = -60$

34. The first term is $\frac{-13}{2}$, the last is $-6 - \frac{1}{2}(13) = \frac{-25}{2}$, so $S_{13} = \frac{13}{2}\left(\frac{-13}{2} + \frac{-25}{2}\right) = \frac{-247}{2}$

36. The first term is 2.8, the last is $2.5 + 0.3(25) = 10$, so $S_{25} = \frac{25}{2}(2.8 + 10) = 160$

38. $a = 3 - 2 = 1, s_{21} = 3(21) - 2 = 61$, so $\displaystyle\sum_{i=1}^{21} 3i - 2 = \frac{21}{2}(1 + 61) = 651$

40. $a = 3 - 2 = 1, s_{20} = 3 - 2(20) = -37$, so $\displaystyle\sum_{j=1}^{20} 3 - 2j = \frac{20}{2}(1 + (-37)) = -360$

42. $a = \frac{4}{3} - 5 = \frac{-11}{3}, s_{14} = \frac{4}{3}(14) - 5 = \frac{41}{3}$, so $\displaystyle\sum_{k=1}^{14} \frac{4}{3}k - 5 = \frac{14}{2}\left(\frac{-11}{3} + \frac{41}{3}\right) = 70$

44. $a = 3, s_{50} = 3(50) = 150$, so $\displaystyle\sum_{n=1}^{50} 3n = \frac{50}{2}(3 + 150) = 3825$

46. a. The first multiple of 7 between 8 and 110 is 14, and the last is 105, so $a = $

14, $d = 7$, and $s_n = 14 + (n - 1)7 = 7n + 7$. Then $105 = 7n + 7$ means $n = 14$,

and $S_n = \frac{14}{2}(14 + 105) = 833$.

b. $\displaystyle\sum_{n=1}^{14} 7n + 7$

48. $a = 63.50 + 2.30 = 65.80$ (this month's bill) and $d = 2.30$, so $s_n = 65.80 + (n - 1)2.30$

$= 2.30n + 63.50$. The first bill is 65.80, the last is $2.30(10) + 63.50 = 86.50$, so his

total bill $= \frac{10}{2}(65.80 + 86.50) = \$761.50$.

50. $a = 20$ minutes, $d = \frac{-2}{3}$ minutes, so $s_n = 20 + (n - 1)\left(\frac{-2}{3}\right) = \frac{-2}{3}n + \frac{62}{3}$, and $s_{30} = $

$\frac{-2}{3}(30) + \frac{62}{3} = \frac{2}{3}$. Then total time $= \frac{30}{2}\left(20 + \frac{2}{3}\right) = 310$ minutes.

52. $a = 3\left(\dfrac{1}{2}\right) = 1.5$ miles, $d = 3(0.2)$, so $s_n = 1.5 + (n - 1)(0.6) = 0.6n + 1.3$. Then $S_n =$

$144 = \dfrac{n}{2}(1.5 + 1.5 + (n - 1)(0.6)) = \dfrac{3}{2}n + 0.3n(n - 1)$, and $1440 = 15n + 3n(n - 1)$, or

$3n^2 + 12n - 1440 = 0$. Thus $3(n + 24)(n - 20) = 0$, and since n cannot be negative,

$n = 20$, and she has been jogging for 20 weeks.

54. The first term, $s_8 = 5(8) + 2 = 42$, the last term, $s_{24} = 5(24) + 2 = 122$. There are

17 terms $(24 - 8 + 1)$, so $S_{17} = \dfrac{17}{2}(42 + 122) = 1394$.

56. The first term, $s_{15} = 15 + 0.3(15) = 19.5$, the last term, $s_{30} = 15 + 0.3(30) = 24$.
There are 16 terms $(30 - 15 + 1)$, so $S_{16} = \dfrac{16}{2}(19.5 + 24) = 348$.

58. first number: a
    second number: $a + d$ $\qquad a + (a + d) + (a + 2d) = 21$
    third number: $a + 2d$ $\qquad a(a + d)(a + 2d) = 231$

$3a + 3d = 21$, so $a + d = 7$, $2a + 2d = 14$, and $a + 2d = 14 - a$. Thus $a(7)(14 - a) = 231$

or $14a - a^2 = 33$, and $a^2 - 14a + 33 = 0$. Then $(a - 3)(a - 11) = 0$, and $a = 3$, $d = 4$, or

$a = 11$, $d = -4$. The values $a = 3$, $d = 4$ yield the numbers 3, 7, 11 as the solution.

($a = 11$, $d = -4$ yield the same three numbers in reverse order).

60. $\displaystyle\sum_{i=1}^{4} pi + q = 10p + 4q$, $\displaystyle\sum_{i=2}^{5} pi + q = 14p + 4q$, so $10p + 4q = 28$

$\qquad\qquad\qquad\qquad\qquad\qquad\qquad\qquad \dfrac{-14p - 4q = -44}{-4p \qquad\quad = -16}$, so $p = 4$, $q = -3$

62. This is an arithmetic progression with $a = 2$, $d = 2$, so $S_n = \dfrac{n}{2}(4 + (n - 1)2) =$

$\dfrac{n}{2}(2n + 2) = \dfrac{2n(n + 1)}{2} = n^2 + n$.

Exercise 12.4

2. This sequence is arithmetic, since each term is obtained by adding -4 to the previous term.

4. This sequence is geometric, since each term is obtained by multiplying the previous term by $\dfrac{1}{2}$.

6. This sequence is neither arithmetic nor geometric.

8. This sequence is geometric, since each term is obtained by multiplying the previous term by -1.

10. This sequence is arithmetic, since each term is obtained by adding $\frac{2}{3}$ to the previous term.

12. This sequence is neither arithmetic nor geometric.

14. $s_1 = -4(3)^{1-1} = -4$, $s_2 = -4(3)^{2-1} = -12$, $s_3 = -4(3)^{3-1} = -36$, $s_4 = -4(3)^{4-1} = -108$

16. $s_1 = 25\left(\frac{4}{5}\right)^{1-1} = 25$, $s_2 = 25\left(\frac{4}{5}\right)^{2-1} = 20$, $s_3 = 25\left(\frac{4}{5}\right)^{3-1} = 16$, $s_4 = 25\left(\frac{4}{5}\right)^{4-1} = \frac{64}{5}$

18. $s_1 = \left(\frac{5}{2}\right)^1 = \frac{5}{2}$, $s_2 = \left(\frac{5}{2}\right)^2 = \frac{25}{4}$, $s_3 = \left(\frac{5}{2}\right)^3 = \frac{125}{8}$, $s_4 = \left(\frac{5}{2}\right)^4 = \frac{625}{16}$

20. $s_1 = 10(0.3)^1 = 3$, $s_2 = 10(0.3)^2 = 0.9$, $s_3 = 10(0.3)^3 = 0.27$, $s_4 = 10(0.3)^4 = 0.081$ ·

22. Next three terms: $32, 64, 128$; $s_n = 4(2)^{n-1}$

24. Next three terms: $\frac{3}{4}, \frac{3}{8}, \frac{3}{16}$; $s_n = 6\left(\frac{3}{2}\right)^{n-1}$

26. Next three terms: $\frac{-27}{2}, \frac{81}{2}, \frac{-243}{2}$; $s_n = \frac{1}{2}(-3)^{n-1}$

28. Next three terms: $\frac{a}{bc^3}, \frac{a}{bc^4}, \frac{a}{bc^5}$; $s_n = \frac{a}{b}\left(\frac{1}{c}\right)^{n-1}$

30.. $a = -3$, $r = \frac{-1}{2}$, so $s_n = -3\left(\frac{-1}{2}\right)^{n-1}$ and $s_8 = -3\left(\frac{-1}{2}\right)^7 = \frac{3}{128}$

32. $a = -81a$, $r = \frac{a}{3}$, so $s_n = -81a\left(\frac{a}{3}\right)^{n-1}$ and $s_9 = -81a\left(\frac{a}{3}\right)^8 = \frac{-a^9}{81}$

34. $s_5 = 1 = a\left(\frac{-1}{2}\right)^{5-1} = \frac{a}{16}$, so $a = 16$

36. $s_7 = \frac{64}{625} = a\left(\frac{2}{5}\right)^{7-1} = \frac{64}{15625}a$, so $a = \frac{15625}{64}\left(\frac{64}{625}\right) = 25$ and $s_n = 25\left(\frac{2}{5}\right)^{n-1}$. Then

$s_2 = 25\left(\frac{2}{5}\right)^1 = 10.$

38. $a = \dfrac{27}{64}$, $r = \dfrac{4}{3}$, so $s_n = \dfrac{27}{64}\left(\dfrac{4}{3}\right)^{n-1}$. Then $\dfrac{64}{27} = \dfrac{27}{64}\left(\dfrac{4}{3}\right)^{n-1}$, and $\left(\dfrac{4}{3}\right)^{n-1} = \dfrac{64}{27}\left(\dfrac{64}{27}\right) =$

$\dfrac{4096}{729} = \left(\dfrac{4}{3}\right)^6$, so $n - 1 = 6$, $n = 7$, and there are 7 terms in the sequence.

40. $a = 12$, $r = 3$, so so $S_8 = \dfrac{12\left(1 - 3^8\right)}{1 - 3} = \dfrac{12(-6560)}{-2} = 39{,}360$

42. $a = 81$, $r = \dfrac{-2}{3}$, so $S_6 = \dfrac{81\left(1 - \left(\dfrac{-2}{3}\right)^6\right)}{1 - \left(\dfrac{-2}{3}\right)} = \dfrac{81\left(1 - \dfrac{64}{729}\right)}{\dfrac{5}{3}} = \dfrac{665}{9}\left(\dfrac{3}{5}\right) = \dfrac{133}{3}$

44. $a = -512$, $r = \dfrac{1}{6}$, so $S_4 = \dfrac{-512\left(1 - \left(\dfrac{1}{6}\right)^4\right)}{1 - \dfrac{1}{6}} = \dfrac{-512\left(1 - \dfrac{1}{1296}\right)}{\dfrac{5}{6}} = \dfrac{-663040}{1296}\left(\dfrac{6}{5}\right) = \dfrac{-16576}{27}$

46. $a = -2$, $r = -2$, so $S_4 = \dfrac{-2\left(1 - (-2)^4\right)}{1 - (-2)} = \dfrac{30}{3} = 10$

48. $a = 2^{-2} = \dfrac{1}{4}$, $r = 2$, so $S_{10} = \dfrac{\dfrac{1}{4}\left(1 - 2^{10}\right)}{1 - 2} = \dfrac{\dfrac{-1023}{4}}{-1} = \dfrac{1023}{4}$

50. $a = \dfrac{1}{4}$, $r = \dfrac{1}{4}$, so $S_5 = \dfrac{\dfrac{1}{4}\left(1 - \left(\dfrac{1}{4}\right)^5\right)}{1 - \dfrac{1}{4}} = \dfrac{\dfrac{1}{4}\left(1 - \dfrac{1}{1024}\right)}{\dfrac{3}{4}} = \dfrac{1023}{4096}\left(\dfrac{4}{3}\right) = \dfrac{341}{1024}$

52. a. $a = 3$, $r = \dfrac{5}{6}$, $s_n = 3\left(\dfrac{5}{6}\right)^{n-1}$, and $s_4 = 3\left(\dfrac{5}{6}\right)^3 = 3\left(\dfrac{125}{216}\right) = \dfrac{125}{72}$ feet.

   b. $S_4 = \dfrac{3\left(1 - \left(\dfrac{5}{6}\right)^4\right)}{1 - \dfrac{5}{6}} = \dfrac{3\left(1 - \dfrac{625}{1296}\right)}{\dfrac{1}{6}} = \dfrac{671}{72}$ feet

54. a = revenue in 1991 = 92% of \$320,000 = \$294,400, $r = 0.92$. For the six years from 1991 to 1996 we have $S_6 = \dfrac{294400\left(1 - 0.92^6\right)}{1 - 0.92} = \dfrac{294400(0.393645)}{0.08} = \$1{,}448{,}613.60$.

56. $a = 0.1$, $r = 1.2$, so $S_{50} = \dfrac{0.1\left(1 - 1.2^{50}\right)}{1 - 1.2} = 4549.72$ seconds or about 75.8 minutes

58. $a = 2, r = \frac{1}{2}, |r| < 1$, so $S_\infty = \dfrac{2}{1 - \frac{1}{2}} = 4$

60. $a = 1, r = \frac{-2}{3}, |r| < 1$, so $S_\infty = \dfrac{1}{1 - \left(\frac{-2}{3}\right)} = \frac{3}{5}$

62. $a = \frac{1}{49}, r = \frac{7}{8}, |r| < 1$, so $S_\infty = \dfrac{\frac{1}{49}}{1 - \frac{7}{8}} = \frac{8}{49}$

64. $a = 2, r = \frac{-3}{4}, |r| < 1$, so $S_\infty = \dfrac{2}{1 - \left(\frac{-3}{4}\right)} = \frac{8}{7}$

66. $a = \frac{-1}{4}, r = \frac{-1}{4}, |r| < 1$, so $S_\infty = \dfrac{\frac{-1}{4}}{1 - \left(\frac{-1}{4}\right)} = \frac{-1}{5}$

68. $0.66666\overline{6} = \frac{6}{10} + \frac{6}{100} + \frac{6}{1000} + \ldots ; a = \frac{6}{10}, r = \frac{1}{10}$, so the fraction $= \dfrac{\frac{6}{10}}{1 - \frac{1}{10}} = \frac{2}{3}$.

70. $0.454545\overline{45} = \frac{45}{100} + \frac{45}{10000} + \frac{45}{1000000} + \ldots ; a = \frac{45}{100}, r = \frac{1}{100}$, so the fraction $=$

    $\dfrac{\frac{45}{100}}{1 - \frac{1}{100}} = \frac{45}{100}\left(\frac{100}{99}\right) = \frac{5}{11}$

72. $3.027\overline{027} = 3 + \frac{27}{1000} + \frac{27}{1000000} + \frac{27}{1000000000} + \ldots ; a = \frac{27}{1000}, r = \frac{1}{1000}$, so the

    fraction $= 3 + \dfrac{\frac{27}{1000}}{1 - \frac{1}{1000}} = 3 + \frac{27}{1000}\left(\frac{1000}{999}\right) = 3\frac{1}{37}$

74. $0.83333\overline{3} = \frac{8}{10} + \frac{3}{100} + \frac{3}{1000} + \frac{3}{10000} + \ldots ; a = \frac{3}{100}, r = \frac{1}{10}$, so the fraction $=$

    $\frac{8}{10} + \dfrac{\frac{3}{100}}{1 - \frac{1}{10}} = \frac{8}{10} + \frac{3}{100}\left(\frac{10}{9}\right) = \frac{8}{10} + \frac{1}{30} = \frac{24}{30} + \frac{1}{30} = \frac{25}{30} = \frac{5}{6}$

76. $a = 12$, $r = 0.9$, so $S_\infty = \dfrac{12}{1 - 0.9} = 120$ inches

78. $S_\infty = 10_{\text{down}} + 6_{\text{up}} + 6_{\text{down}} + 3.6_{\text{up}} + 3.6_{\text{down}} + \ldots = 10 + 12 + 7.2 + \ldots$; $a = 12$,

$r = 0.6$, so $S_\infty = 10 + \dfrac{12}{1 - 0.6} = 10 + 30 = 40$ feet

Exercise 12.5

2. $(x + y)^4 = \displaystyle\sum_{k=0}^{4} \binom{4}{k} x^{4-k} y^k = \binom{4}{0} x^4 y^0 + \binom{4}{1} x^3 y^1 + \binom{4}{2} x^2 y^2 + \binom{4}{3} x^1 y^3 + \binom{4}{4} x^0 y^4$

4. $(2x - y)^7 = \displaystyle\sum_{k=0}^{7} \binom{7}{k} (2x)^{7-k}(-y)^k = \binom{7}{0}(2x)^7(-y)^0 + \binom{7}{1}(2x)^6(-y)^1 + \binom{7}{2}(2x)^5(-y)^2 +$

$\binom{7}{3}(2x)^4(-y)^3 + \binom{7}{4}(2x)^3(-y)^4 + \binom{7}{5}(2x)^2(-y)^5 + \binom{7}{6}(2x)^1(-y)^6 + \binom{7}{7}(2x)^0(-y)^7$

6. $(1 - y^2)^5 = \displaystyle\sum_{k=0}^{5} \binom{5}{k} 1^{5-k}(-y^2)^k = \binom{5}{0}1^5(-y^2)^0 + \binom{5}{1}1^4(-y^2)^1 + \binom{5}{2}1^3(-y^2)^2 +$

$\binom{5}{3}1^2(-y^2)^3 + \binom{5}{4}1^1(-y^2)^4 + \binom{5}{5}1^0(-y^2)^5$

8. $(2x + y)^4 = \binom{4}{0}(2x)^4 y^0 + \binom{4}{1}(2x)^3 y^1 + \binom{4}{2}(2x)^2 y^2 + \binom{4}{3}(2x)^1 y^3 + \binom{4}{4}(2x)^0 y^4 =$

$1(16x^4) + 4(8x^3)y + 6(4x^2)y^2 + 4(2x)y^3 + 1y^4 = 16x^4 + 32x^3 y + 24x^2 y^2 + 8xy^3 + y^4$

10. $(2x - 1)^5 = \binom{5}{0}(2x)^5(-1)^0 + \binom{5}{1}(2x)^4(-1)^1 + \binom{5}{2}(2x)^3(-1)^2 + \binom{5}{3}(2x)^2(-1)^3 + \binom{5}{4}(2x)^1(-1)^4$

$+ \binom{5}{5}(2x)^0(-1)^5 = 1(32x^5) - 5(16x^4) + 10(8x^3) - 10(4x^2) + 5(2x) - 1 =$

$32x^5 - 80x^4 + 80x^3 - 40x^2 + 10x - 1$

12. $\left(\dfrac{x}{3} + 3\right)^5 = \binom{5}{0}\left(\dfrac{x}{3}\right)^5 3^0 + \binom{5}{1}\left(\dfrac{x}{3}\right)^4 3^1 + \binom{5}{2}\left(\dfrac{x}{3}\right)^3 3^2 + \binom{5}{3}\left(\dfrac{x}{3}\right)^2 3^3 + \binom{5}{4}\left(\dfrac{x}{3}\right)^1 3^4 + \binom{5}{5}\left(\dfrac{x}{3}\right)^0 3^5 =$

$\left(\dfrac{x^5}{243}\right) + 5\left(\dfrac{x^4}{27}\right) + 10\left(\dfrac{x^3}{3}\right) + 10(3x^2) + 5(27x) + 243 = \dfrac{x^5}{243} + \dfrac{5x^4}{27} + \dfrac{10x^3}{3} + 30x^2 + 135x + 243$

14. $\left(\dfrac{2}{3} - a^2\right)^4 = \binom{4}{0}\left(\dfrac{2}{3}\right)^4(-a^2)^0 + \binom{4}{1}\left(\dfrac{2}{3}\right)^3(-a^2)^1 + \binom{4}{2}\left(\dfrac{2}{3}\right)^2(-a^2)^2 + \binom{4}{3}\left(\dfrac{2}{3}\right)^1(-a^2)^3 + \binom{4}{4}\left(\dfrac{2}{3}\right)^0(-a^2)^4$

$= \left(\dfrac{16}{81}\right) - 4a^2\left(\dfrac{8}{27}\right) + 6a^4\left(\dfrac{4}{9}\right) - 4a^6\left(\dfrac{2}{3}\right) + a^8 = \dfrac{16}{81} - \dfrac{32}{27}a^2 + \dfrac{8}{3}a^4 - \dfrac{8}{3}a^6 + a^8$

16. $(3n)! = 12! = 12 \cdot 11 \cdot 10 \cdot 9 \cdot 8 \cdot 7 \cdot 6 \cdot 5 \cdot 4 \cdot 3 \cdot 2 \cdot 1 = 479{,}001{,}600$

18. $3n! = 3(4!) = 3(4 \cdot 3 \cdot 2 \cdot 1) = 3 \cdot 24 = 72$ 20. $2n(2n-1)! = 4(3!) = 4(3 \cdot 2 \cdot 1) = 24$

22. $7! = 7 \cdot 6 \cdot 5 \cdot 4 \cdot 3 \cdot 2 \cdot 1 = 5040$ 24. $\dfrac{12!}{11!} = \dfrac{12 \cdot 11!}{1 \cdot 11!} = 12$

26. $\dfrac{12!8!}{16!} = \dfrac{12!(8 \cdot 7 \cdot 6 \cdot 5 \cdot 4 \cdot 3 \cdot 2 \cdot 1)}{16 \cdot 15 \cdot 14 \cdot 13(12!)} = \dfrac{8 \cdot 7 \cdot 6 \cdot 5 \cdot 4 \cdot 3 \cdot 2 \cdot 1}{16 \cdot 15 \cdot 14 \cdot 13} = \dfrac{12}{13}$

28. $\dfrac{10!}{4!(10-4)!} = \dfrac{10 \cdot 9 \cdot 8 \cdot 7(6!)}{4 \cdot 3 \cdot 2 \cdot 1(6!)} = \dfrac{10 \cdot 9 \cdot 8 \cdot 7}{4 \cdot 3 \cdot 2 \cdot 1} = 210$ 30. $5!$

32. $7 = \dfrac{7 \cdot 6!}{6!} = \dfrac{7!}{6!}$ 34. $\dfrac{28 \cdot 27 \cdot 26 \cdot 25 \cdot 24(23!)}{23!} = \dfrac{28!}{23!}$

36. $(n+4)! = (n+4)(n+3)(n+2) \cdot \ldots \cdot 3 \cdot 2 \cdot 1$

38. $3n! = 3n(n-1)(n-2) \cdot \ldots \cdot 3 \cdot 2 \cdot 1$

40. $(3n-2)! = (3n-2)(3n-3)(3n-4) \cdot \ldots \cdot 3 \cdot 2 \cdot 1$

42. $\binom{8}{5} = \dfrac{8!}{5!(8-5)!} = \dfrac{8 \cdot 7 \cdot 6(5!)}{(5!)3 \cdot 2 \cdot 1} = 56$ 44. $\binom{13}{4} = \dfrac{13!}{4!9!} = \dfrac{13 \cdot 12 \cdot 11 \cdot 10}{4 \cdot 3 \cdot 2 \cdot 1} = 715$

46. $\binom{18}{16} = \dfrac{18!}{16!2!} = \dfrac{18 \cdot 17}{2 \cdot 1} = 153$

48. The term is $\binom{15}{3}x^{12}(-y)^3$, so the coefficient is $-\dfrac{15!}{3!12!} = -\dfrac{15 \cdot 14 \cdot 13}{3 \cdot 2 \cdot 1} = -455$

50. The term is $\binom{12}{4}(2a)^8(-b)^4$, so the coefficient is $(256)\dfrac{12!}{4!8!} = 256\left(\dfrac{12 \cdot 11 \cdot 10 \cdot 9}{4 \cdot 3 \cdot 2 \cdot 1}\right) =$

$256(495) = 126{,}720$

52. The term is $\binom{8}{2}\left(\dfrac{x}{2}\right)^6 2^2$, so the coefficient is $\dfrac{4}{64}\left(\dfrac{8!}{2!6!}\right) = \dfrac{1}{16}\left(\dfrac{8 \cdot 7}{2 \cdot 1}\right) = \dfrac{7}{4}$

54. The fifth term is $\binom{12}{4} x^8 2^4 = \dfrac{12!}{4!8!} x^8(16) = (16)\dfrac{12 \cdot 11 \cdot 10 \cdot 9}{4 \cdot 3 \cdot 2 \cdot 1} x^8 = 16(495)x^8 = 7{,}920x^8$

56. The seventh term is $\binom{9}{6}(a^3)^3(-b)^6 = \dfrac{9!}{6!3!} a^9 b^6 = \dfrac{9 \cdot 8 \cdot 7}{3 \cdot 2 \cdot 1} a^9 b^6 = 84a^9 b^6$

58. The fourth term is $\binom{8}{3} 1^5 \left(\dfrac{1}{2}\right)^3 = \dfrac{8!}{3!5!} (1)\left(\dfrac{1}{8}\right) = \dfrac{8 \cdot 7 \cdot 6}{3 \cdot 2 \cdot 1}\left(\dfrac{1}{8}\right) = 7$

60. The eighth term is $\binom{10}{7}\left(\dfrac{x}{2}\right)^3 4^7 = \dfrac{10!}{7!3!}\left(\dfrac{x^3}{8}\right) 16{,}384 = \dfrac{120}{8} x^3(16{,}384) = 245{,}760x^3$

62. $\left(b - \dfrac{a}{3}\right)^7 = b^7 - \dfrac{7}{3} b^6 a + \dfrac{7}{3} b^5 a^2 - \dfrac{35}{27} b^4 a^3 + \ldots$

64. $(x^2 + 2y)^9 = x^{18} + 18x^{16}y + 144x^{14}y^2 + 672x^{12}y^3 + \ldots$

66. $\left(\dfrac{1}{x} - 4\right)^6 = \dfrac{1}{x^6} - \dfrac{24}{x^5} + \dfrac{240}{x^4} - \dfrac{1280}{x^3} + \ldots$

68. $(2 + \sqrt{y})^{11} = 2{,}048 + 11{,}264\sqrt{y} + 28{,}160y + 42{,}240y\sqrt{y} + \ldots$

70. $(1.01)^{15} \approx (1)^{15} + 15(1)^{14}(0.01) + 105(1)^{13}(0.01)^2 = 1 + 0.15 + 0.0105 = 1.1605 \approx 1.16$

72. $(0.95)^8 = (1 - 0.05)^8 \approx (1)^8 + 8(1)^7(-0.05)^1 + 28(1)^6(-0.05)^2 = 1 - 0.4 + 0.07 = 0.67$

74. a. Here $n = \dfrac{1}{2}$ and $x = 0.02$, so $\sqrt{1.02} \approx 1^{1/2}(0.02)^0 + \dfrac{1}{2}(1)^{-1/2}(0.02)^1 +$

$\left(\dfrac{-1}{4}\right)(1)^{-3/2}(0.02)^2 = 1 + 0.01 - 0.0001 = 1.0099$

   b. Here $n = \dfrac{1}{2}$ and $x = -0.01$, so $\sqrt{0.99} \approx 1^{1/2}(-0.01)^0 + \dfrac{1}{2}(1)^{-1/2}(-0.01)^1 +$

$\left(\dfrac{-1}{4}\right)(1)^{-3/2}(-0.01)^2 = 1 - 0.005 - 0.000025 = 0.994975$